Lecture Notes in Physics

Edited by J. Ehlers, München, K. Hepp, Zürich,
H. A. Weidenmüller, Heidelberg, and J. Zittartz, Köln
Managing Editor: W. Beiglböck, Heidelberg

47

Padé Approximants Method and Its Applications to Mechanics

Edited by H. Cabannes

Springer-Verlag
Berlin Heidelberg GmbH 1976

Editor
Prof. Henri Cabannes
Université Pierre et Marie Curie
Mécanique Théorique
Tour 66, 4, Place Jussieu
75005 Paris/France

Library of Congress Cataloging in Publication Data
Main entry under title:

Padé approximants method and its applications to mechanics.

(Lecture notes in physics ; 47)
Bibliography: p.
Includes index.
1. Padé approximant. 2. Fluid mechanics.
I. Cabannes, Henri. II. Series.
QC20.7.P3P35 532'.01'515 75-46504

ISBN 978-3-540-07614-8 ISBN 978-3-540-38132-7 (eBook)
DOI 10.1007/978-3-540-38132-7

Originally published by Springer-Verlag Berlin Heidelberg New York in 1976

Henri Padé (1863 - 1953)

Henri Eugène Padé was born in Abbeville (France) on Dec. 17th 1863. Admitted in 1883 to the Ecole Normale Supérieure, he left it in 1886 with the highest teacher's degree (Agrégation) in Mathematics. After teaching at the classical secondary school in Limoges, Carcassonne and Montpellier, he was granted a leave in 1889 in order to study in Germany, first in Leipzig, then in Göttingen. On June 21st 1892, before the University of Paris, he defended his doctorate thesis on the approximate representation of a function by rational fractions.

Henri Padé was appointed lecturer at the Faculty of Sciences of Lille in 1897, Professor of Rational and Applied Mechanics at the Faculty of Sciences of Poitiers in 1902 and Professor of Mechanics at the Faculty of Bordeaux in 1903. In 1906, he was elected Dean of the Faculty of Sciences of Bordeaux and became Laureate with the major prize for mathematical sciences awarted by the Academy of Sciences on the report of Emile Picard. In 1908 he was named Rector of the Academy of Besançon, he was then the youngest rector in France. In 1917 he became rector of the Academy of Dijon and in 1923 rector of the Academy of Aix-Marseilles, an office he kept until he retired in 1934.

Henri Padé died in 1953 at the age of 89.

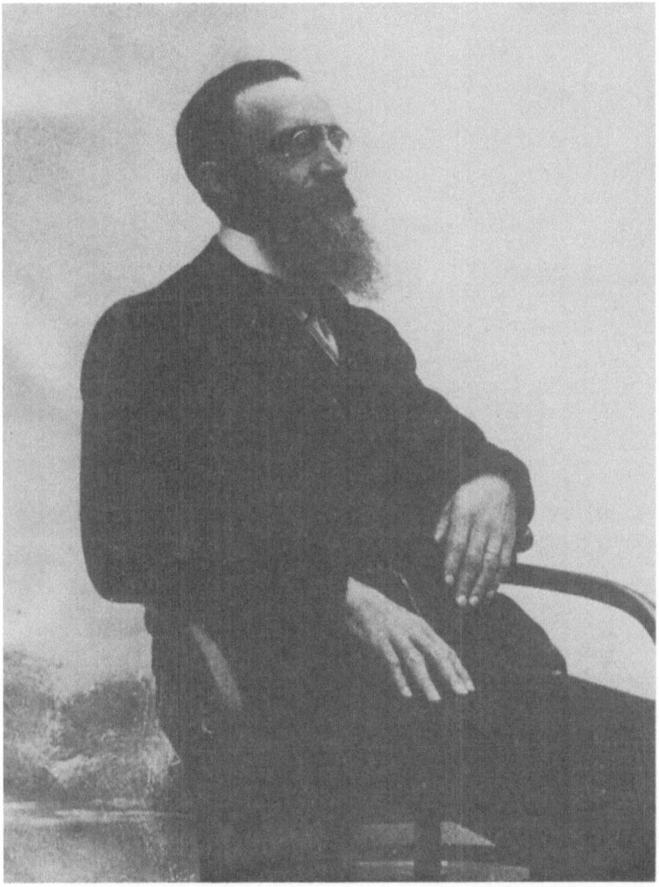

Henri Padé (1863 - 1953)

MINISTÈRE

DE

L'INSTRUCTION PUBLIQUE

DIRECTION

de l'Enseignement supérieur.

2 e BUREAU

Réponse à la dépêche en date

du

OBJET :

Demande d'autorisation d'absence.

RÉPUBLIQUE FRANCAISE

ACADÉMIE DE BESANCON

Besançon, le juillet 1916.

Le Recteur de l'Académie de Besançon

à Monsieur le Ministre de l'Instruction Publique.

J'ai l'honneur de vous informer que je compte prendre quelque repos pendant les vacances.

Mon absence durerait au maximum six semaines, du 6 aout au 15 septembre, pendant lesquelles je ne quitterais pas le ressort et me tiendcais pret à reprendre mon service si les circonstances l'exigeaient. J'aurai d'ailleurs soin de vous faire connaitre mon adresse exacte pendant mon deplacement.

Le service serait, comme d'habitude, assuré par mes bureaux. Qui sont [illegible] au courant des affaires, pouvant [illegible] immédiates.

Jo vous serais reconnaissant de vouloir bien me faire savelr si vous approuvez mon projet.

Je vous prie de vouloir m'autoriser à ~~m'absenter~~ pendant six semaines du 6 août au 15 septembre.

Le Recteur,
Président du conseil de l'Université,

[signature]

Padé.

11 février 1913

R à M. (II, 1)   11 fevr 1913

J'ai l'honneur de vous accuser réception de votre dépêche en date du ~~10~~ 12 ~~janvier 1913~~ janvier 1914, par laquelle vous voulez bien m'informer que, par arrêté de ce jour, vous m'avez ~~désigné pour~~ nommé vice-président du jury chargé, sous la direction de M. Appell, membre de l'Institut, d'examiner, en ~~1913~~ 1914, les aspirants à l'agrégation de l'ens^t second. de j. f. (ordre des sciences, — section des sc. math.).

~~Je vous prie d'agréer mes remerciements pour cette marque de haute bienveillance, je vous prie, de la recevoir, l'hommage de mon respect~~

Je vous prie de vouloir bien ~~recevoir~~ agréer mes vifs remerciements pour cette marque de haute bienveillance et l'hommage de mes sentiments respectueux dévouement

R.

Editor's Preface

In 1892, in the Scientific Transactions of the Ecole Normale Supérieure in Paris, the french mathematician Henri Padé published an article concerning the approximate representation of a function by rational fractions. Three-quarters of a century later, the advent of arithmetical computers led scientists to consider various methods of representing functions, especially rapidly converging functions.

In a paper published in 1955, D. Shanks showed the advantages of padé'smethod, which makes it possible to deduce from any converging or diverging series of powers a table of rational approximations of the functions represented by these series. Since then many physicists and applied mathematicians have studied the representation of functions by means of rational fractions, the main object being to obtain ever faster computing algorithms. The most celebrated and now classic example is for calculating the sum of the following series:

$$S = 4 - \frac{4}{3} + \frac{4}{5} - \frac{4}{7} \dots \qquad + (-)^n \frac{4}{2n+1} + \dots \tag{1}$$

The sum S is equal to π. But to obtain from series (1) — which is a very poor algorithm — the value of π correct to eight digits, the first term which is neglected must be less than 10^{-8} in absolute value, which means that we must consider 20 million terms. If however one applies Padé's [1,1] transformation repeatedly to the sum S_n of the first n terms, that is if one associates with the sequence S_n the new sequence

$$\Sigma_n = \frac{S_{n-1} S_{n+1} - S_n^2}{S_{n-1} + S_{n+1} - 2 S_n} , \tag{2}$$

then to obtain the same accuracy it will suffice to consider only the first nine terms of series (1) and to apply the transformation (2) five times.

The underlying principle of Padé's method is particularly simple. Given a power series

$$\sum_{k=0}^{\infty} c_k \; x^k , \tag{3}$$

Padé proposed finding the closest approximation to the sum, by defining a rational fraction $P_m(x) \,/\, Q_n(x)$, with

$$(4) \qquad P_m(x) = \sum_{k=0}^{m} a_k \; x^k , \qquad\qquad Q_n(x) = 1 + \sum_{k=1}^{n} b_k \; x^k ,$$

in which the numerator and the denominator are polynomials of degrees m and n respectively. These polynomials $P_m(x)$ and $Q_n(x)$ are determinated from the identity

$$(5) \qquad Q_n(x) \sum_{k=0}^{\infty} c_k \; x^k \; - \; P_m(x) \; = x^{m+n+1} \sum_{k=0}^{\infty} \gamma_k \; x^k$$

This identity leads to m+n+1 linear equations from which the unknown m+n+1 values a_k and b_k ($b_0 = 1$) can be determined. Padé obtained the following result:

$$(6) \; A \, P_m(x) \; = \begin{vmatrix} c_{m-n+1} & c_{m-n+2} & \cdots & c_{m+1} \\ \vdots & & & \\ c_m & c_{m+1} & \cdots & c_{m+n} \\ \sum_{j=0}^{m} c_{j-n} \; x^j & \sum_{j=0}^{m} c_{j-n+1} \; x^j & \cdots & \sum_{j=0}^{m} c_j \; x^j \end{vmatrix}$$

$$(7) \; A \, Q_n(x) \; = \begin{vmatrix} c_{m-n+1} & c_{m-n+2} & \cdots & c_{m+1} \\ \vdots & & & \\ c_m & c_{m+1} & \cdots & c_{m+n} \\ x^n & x^{n-1} & \cdots & 1 \end{vmatrix}$$

In these two formulae, it is necessary to take $c_j = 0$ when j is negative, and A is the minor obtained by eliminating the last line and the last column.

Padé's rational approximations are widely used in computer calculations because they are generally more efficient than polynomial approximations. They almost halve the number of operations required. These approximations are particularly convenient when one takes m = n (defining the diagonal of the Padé table). For in this case the coefficients γ_k of the right-hand side of (5) usually decrease so quickly that the first term $\gamma_0 \; x^{2n+1}$, divided by $Q_n(x)$, constitues an excellent approximate value for the absolute error introduced if one uses $P_m(x) \; / \; Q_n(x)$ instead of

$\sum_{k=0}^{\infty} c_k \; x^k$. For $|x| < 1$, the value of $Q_n(x)$ differs very little from unity because the b_k coefficients usually decrease very rapidly. It will therefore suffice to calculate γ_0, whence

(8) $$\gamma_0 = \frac{\Delta_n}{\delta_n}$$

$$\Delta_n = \begin{vmatrix} c_1 & c_2 & \cdots & c_{n+1} \\ c_2 & c_3 & \cdots & c_{n+2} \\ \vdots & & & \\ c_{n+1} & c_{n+2} & \cdots & c_{2n+1} \end{vmatrix}$$

δ_n is the minor obtained by eliminating the last line and the last column.

In an excellent article written for the collective work entitled "Mathematical Methods for Digital Computers" (published by John Wiley), Kogbetliantz shows that, irrespective of accuracy required, the Padé method is the best method for constructing programs for calculating sin x and cos x:

(9) $$\cos x = \sum_{k=0}^{\infty} \frac{(-)^k}{(2k)!} \; z^k \qquad \text{with } z = x^2$$

Taking m = n, we have

$$\frac{P_1}{Q_1} = \frac{12 - 5\,x^2}{12 + x^2}\,, \qquad \gamma_0 = \frac{3}{2}\,\frac{1}{6!}$$

$$\frac{P_2}{Q_2} = \frac{15120 - 6900\,x^2 + 313\,x^4}{15120 + 660\,x^2 + 13\,x^4}\,, \qquad \gamma_0 = -\frac{59}{42}\,\frac{1}{10!}$$

As we can see on the figure, the approximation P_2 / Q_2 which is constructed with the five first terms of the series (1) is better than the sum S_5 of those terms:

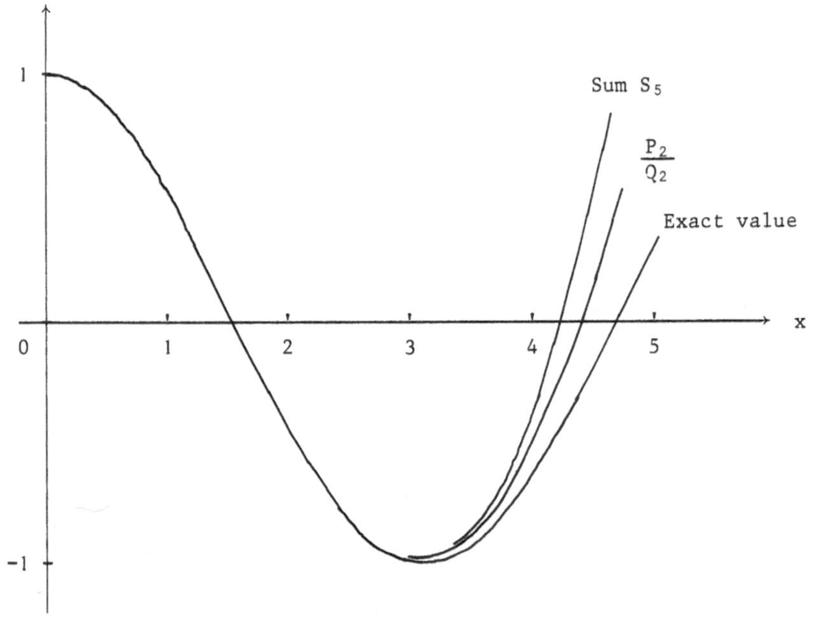

Approximations of cos x

For m = n = 3, we obtain

$$\frac{P_3}{Q_3} = \frac{1 + a_1 x + a_2 x^2 + a_3 x^3}{1 + b_1 x + b_2 x^2 + b_3 x^3}$$

with

$$a_1 = -\frac{3665}{7788} \qquad b_1 = \frac{229}{7788}$$

$$a_2 = \frac{711}{25960} \qquad b_2 = \frac{1}{2360}$$

$$a_3 = -\frac{2\,923}{7\,850\,304} \qquad b_3 = \frac{127}{39\,251\,520}$$

and $\gamma_0 = \frac{1407}{2596} \frac{1}{14!} < 7 . 10^{-12}$

Practically the approximation P_3 / Q_3 is put into the form of a continued fraction

$$(10) \qquad \frac{P_3}{Q_3} = C_0 + \cfrac{C_1}{x^2 + B_1 + \cfrac{C_2}{x^2 + B_2 + \cfrac{C_3}{x^2 + B_3}}}$$

$B_1 = 92.70474$	$C_1 = 23535.603$
$B_2 = - 3.50476$	$C_2 = 4017.0378$
$B_3 = 41.76065$	$C_3 = 1670.6950$

Since $C_0 = - \frac{14\ 615}{127}$ one is led to calculate a small number from the difference between two large numbers, resulting in an inaccurate result. This drawback can be circumwented by substituting ξ z for z, the parameter ξ being chosen so that $C_0(\xi)$ is small. Finally the value of cos x is obtained over the interval $(0, \frac{\pi}{3})$ to ten significant digits from this approximation; we use a rational fraction put into the form of a continued fraction, performing only four multiplications and using seven constants. Calculating $P_3(z) / Q_3(z)$ as a classical fraction would require eight operations.

This short exposition of the Padé method explains why it has come to be widely used since the advent of computers. The Padé method is currently being studied and used by three categories of scientists: numerical analysis specialists, theoretical physicists and specialists in fluid mechanics. After the European symposium on Mechanics organized at the Toulon University Center in 1975, it was felt that it might be useful to gather together several articles devoted either to the fundamentals of the method or to its applications in mechanics. This is how the present volume was born, and I am particularly indebted to Professor Beiglböck for having kindly included it in the Lecture Notes in Physics series, and to Springer-Verlag for having published it so quickly.

Henri Cabannes

January 1976

Contents

Part I

Mathematical Theory of the Padé Approximants Method

THE LINEAR, FUNCTIONAL EQUATION APPROACH TO THE PROBLEM OF THE CONVERGENCE OF PADÉ APPROXIMANTS*

GEORGE A. BAKER, JR.

Applied Mathematics Department
Brookhaven National Laboratory
Upton, N.Y. 11973

ABSTRACT

The Padé approximant problem is related to a (not necessarily orthogonal) projection of a linear functional equation of the Fredholm type. If the kernel is of trace class and its upper Hessenberg form is tridiagonal (this class includes Hermitian operators), then we prove that not only do the diagonal Padé approximants converge, but so do their numerators and denominators separately. The generalization of these results to C_p classes of compact operators is given. For kernels which are not only compact, but also satisfy an additional mild restriction, a pointwise convergence theorem is proven. The application of these results to quantum scattering theory is indicated.

---

*Work performed under the auspices of the U.S. ERDA.

Considerable progress in the study of the convergence of Padé approximants can be made, I think, by the use of the techniques of functional equations. What I will report here is probably just a beginning, and is drawn in part from a previous paper.[1] First we will review the known relation of Padé approximants to linear functional equations, then we review the properties of some special classes of compact operators, and give convergence results for these classes. Finally we indicate how these results lead to convergence of Padé approximants to the partial wave scattering amplitudes in certain quantum mechanical scattering problems.

PROJECTIONS IN THE CINI-FUBINI SUBSPACE

Suppose we consider the functional equation

$$f = g + \lambda A f \tag{1}$$

where f, g, and h belong to some Hilbert space $\mathcal{H}$, and A is a linear operator whose properties are yet to be defined. We also introduce the associated sets of elements

$$\varphi_i = A^{i-1} g, \quad \varphi_i' = (A^\dagger)^{i-1} h \quad i = 1, 2, \ldots, \tag{2}$$

where $A^\dagger$ is the Hermitian conjugate operator to A. We need as well the N × N matrix

$$R_{i,j} = (\varphi_i', \varphi_j) = (h, A^{i+j-2} g) \tag{3}$$

defined in terms of the inner products of the φ_j and φ_i'. We are now in a position to define our projection operator onto the Cini-Fubini subspace [2]

$$P_N = \sum_{i,j=1}^{N} \varphi_i (R^{-1})_{ij} (\varphi_j', \ ), \tag{4}$$

provided $\det|R| \neq 0$. (It can be shown [3] that there exists an infinite number of such N's.) The operator P_N is a projection on $\mathcal{S}_N$ from $\mathcal{S}_N'$. (The spaces $\mathcal{S}_N$ and $\mathcal{S}_N'$ are respectively those spaces spanned by φ_i and φ_i' for $i = 1,\ldots,N$.) It has the properties

$$P_N P_M = P_M P_N = P_M, \quad M \leq N. \tag{5}$$

However, it may not be an orthogonal projection. If it is not, then its norm $\|P_N\|$ will be greater than unity! We show in fig. 1 the projection on non-orthogonal directions. It is clear in this figure that the "length" $(a^2 + b^2)^{\frac{1}{2}}$ of the projection can be greater than the length of the original vector.

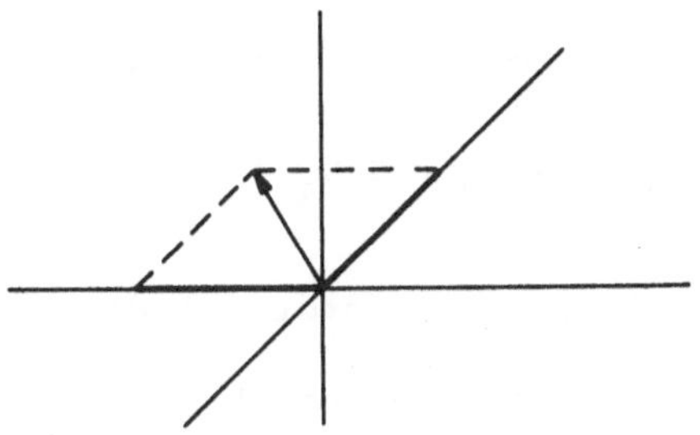

Fig. 1

Projection on non-orthogonal directions
(heavy lines) of a vector (arrow)

With this machinery, let us consider the truncated equation

$$f_N = g + \lambda \, P_N \, A \, P_N \, f_N. \tag{6}$$

By the properties of P_N, we expect a solution of the form

$$f_N = \sum_{j=1}^{N} a_j \, \varphi_j. \tag{7}$$

The substitution of (7) leads to the solution

$$f_N = \left(\sum_{i=1}^{N} \sum_{j=1}^{N} \varphi_i V_{ij} \omega_{j-1}\right) / \det|U_{ij}|, \tag{8}$$

where

$$U_{ij} = \omega_{i+j-2} - \lambda \omega_{i+j-1}$$

$$V_{ij} = i, \underline{jth} \text{ minor of } (U_{k\ell}). \tag{9}$$

Then

$$(h, f_N) = \sum_{i=1}^{N} \sum_{j=1}^{N} \omega_{i-1} V_{ij} \omega_{j-1} / \det|U_{ij}|. \tag{10}$$

However this formula is Nuttall's compact form [4,5] for the [N-1/N] Padé approximant to $h(\lambda)$ defined by the Liouville-Neumann series

$$\begin{aligned} f &= g + \lambda Ag + \lambda^2 A^2 g + \lambda^3 A^3 g + \dots, \\ h(\lambda) &= (h, f) = \omega_0 + \lambda \omega_1 + \lambda^2 \omega_2 + \dots \end{aligned} \tag{11}$$

Thus by use of the projection operator (4) we generate directly the Padé approximants as the solutions of the truncated equations.

RESULTS FOR THE TRACE CLASS OF COMPACT OPERATORS

Here we will assume that P_N is orthogonal. What does this restriction imply about the linear operator A? To examine this question we use a basis e_i determined by the φ_i so that the e_i are orthonormal and the first N of them span $\mathcal{S}_N$. Using this basis, we see that A is of the form

$$\begin{pmatrix} x & x & x & x & \dots \\ x & x & x & x & \dots \\ 0 & x & x & x & \dots \\ 0 & 0 & x & x & \dots \\ 0 & 0 & 0 & x & \dots \\ \vdots & \vdots & \vdots & \vdots & \ddots \end{pmatrix}, \tag{12}$$

that is, upper right triangular plus one subdiagonal. This form is called the upper Hessenberg form. Now if $A = A^{\dagger}$, this Hermiticity condition implies at once that A is tridiagonal. Therefore if we choose h of eq. (2) equal to g, then the φ_i' define the same spaces as the φ_i, and P_N is orthogonal. But this conclusion depends only on the tri-diagonality of A which is more general than Hermiticity. The restriction $A = A^{\dagger}$ would yield only generalized series of Stieltjes, however our results extend them.

Now let us define trace class operators. If the operator is trace class and A is tridiagonal in its upper Hessenberg form, then it turns out that the numerator and denominator converge separately. In order to define the trace class of compact operators we introduce the non-negative, definite, Hermitian operator

$$T = A^{\dagger} A, \tag{13}$$

which has the eigenvalues and eigenvectors

$$T \psi_i = \alpha_i^2 \psi_i. \tag{14}$$

The trace norm of A is then

$$\|A\|_1 = \Sigma_i \alpha_i, \tag{15}$$

which is something like $\mathrm{Tr}(|A|)$, and the trace class consists of those operators with a finite trace norm. Standard theory [6] insures that for an operator of trace class we can define

$$D(\lambda) = \det|I + \lambda A| = \lim_{N\to\infty} D_N(\lambda), \tag{16}$$

where the D_N are the determinants in a sequence of subspaces. The resulting $D(\lambda)$ is an entire function of λ. We note at this point that some condition related to the trace of A is required as the Padé denominator must go to

$$1 - \lambda \sum_i \lambda_i^{-1} + O(\lambda^2), \tag{17}$$

and, if

$$\Sigma \lambda_i^{-1} = \mathrm{Tr}(A) = \infty, \tag{18}$$

then the denominator can't possibly converge separately.

In order to see the convergence in this case, it is convenient to construct the Fredholm solution to the truncated equations. To this end we introduce an orthonormal basis $\chi_i, i=1,\ldots,N$ spanning $\mathcal{S}_N = \mathcal{S}'_N$. Then

$$A_{ij} = (\chi_i, A\chi_j), \; f_N = \Sigma\, b_i \chi_i, \tag{19}$$

$$b_j = (\chi_j, g) + \lambda \sum_{k=1}^{N} A_{jk} b_k. \tag{20}$$

If

$$D_N(\lambda) = \det_N |\delta_{ij} - \lambda A_{ij}| \neq 0, \tag{21}$$

then the Fredholm solution is given by

$$f_N = g + \lambda \Sigma \chi_j D_{N,jk}(\lambda)(\chi_k, g)/D_N(\lambda), \tag{22}$$

where

$$D_{N,jk}(\lambda) = A_{jk} - \frac{\lambda}{1!} \sum_{\ell=1}^{N} \begin{vmatrix} A_{jk} & A_{j\ell} \\ A_{\ell k} & A_{\ell\ell} \end{vmatrix} + \frac{\lambda^2}{2!} \Sigma\Sigma | \quad | + \ldots, \tag{23}$$

and are automatically polynomials of degree at most N - 1 in λ. If we choose the χ_i to be eigenfunctions of

$$T_N = (P_N A P_N)^\dagger (P_N A P_N), \tag{24}$$

and use Hadamard's determinant inequality

$$\left|\det \begin{vmatrix} a_{11} & \cdots & a_{1n} \\ \vdots & \ddots & \vdots \\ a_{n1} & \cdots & a_{nn} \end{vmatrix}\right| \leq \prod_{i=1}^{n} \left(\sum_{j=1}^{N} |a_{ij}|^2\right)^{\frac{1}{2}}, \tag{25}$$

then we can show [1] for any z in $\mathcal{H}$, of unit length

$$\left\{ \sum_{j=1}^{N} \left| \sum_{k=1}^{N} D_{N,jk}(\lambda)\ z_k \right|^2 \right\}^{\frac{1}{2}} \leq \|A\|_2 \ (1$$

$$+ \frac{|\lambda|}{1!}\ 2^{2/2}\ \|A\|_1 + \frac{|\lambda|}{2!}\ 3^{3/2}\ \|A\|_1^2 + \ldots). \tag{26}$$

Thus the operators that give the numerator in the Fredholm solution are a sequence of uniformly bounded operators in N for all λ provided the series in (26) converges. As $\|A\|_1$ is finite the series does and $\|A\|_1$ finite makes $\|A\|_2$ automatically finite also. Standard arguments then insure the convergence of the numerators and eq. (16) gives that of the denominators. Hence we have the separate convergence of the numerator and denominator of the Padé approximants to the ratio of the two entire functions given by the Fredholm solution.

C_p CLASSES OF COMPACT OPERATORS

We may define larger classes of operators than the trace class and obtain special convergence results for them. We say a compact operator belongs to class C_p if, using (13) and (14),

$$\|A\|_p = [\sum_i \alpha_i^p]^{1/p} < \infty. \tag{27}$$

The trace class is C_1 and these classes have the properties that $C_{p-1} \subset C_p$. We use the convenient notation, due to Nuttall,

$$[\sum_{i=0}^{\infty} F_i x]^p = \sum_{i=0}^{p} F_i x^i \tag{28}$$

for any formal power series. The results for operators of class C_p are based on rewriting the Padé approximants as

$$[N-1/N] = \frac{P_N(\lambda)}{Q_N(\lambda)} = \frac{P_N(\lambda)\ \exp\{-[\ln Q_N(\lambda)]^{p-1}\}}{Q_N(\lambda)\ \exp\{-[\ln Q_N(\lambda)]^{p-1}\}}\ . \tag{29}$$

By applying arguments parallel to those of Dunford and Schwartz [6], we can show for P_N orthogonal and A in C_p that if we define

$$D_{N,p}(\lambda) = \det_N(\delta_{ij} - \lambda A_{ij})\ \exp\{-[\ln \det_N(\delta_{ij} - \lambda A_{ij})]^{p-1}\} \tag{30}$$

then

$$\begin{aligned} \lim_{N\to\infty} D_{N,p}(\lambda) &= \text{entire function of } \lambda \\ &\le \exp(\Gamma\|A\|_p^p \lambda^p), \end{aligned} \tag{31}$$

and in the N × N truncated space $\mathcal{S}_N = \mathcal{S}_N'$

$$\begin{aligned} \lim_{N\to\infty} \|D_{N,p}(\lambda)\{(\delta_{ij} - \lambda A_{ij})^{-1} - \delta_{ij} - \lambda A - \ldots - \lambda^{p-2} A^{p-2}\}\|_{\frac{p}{p-1}} \\ \le \exp[\Gamma(\lambda^p\|A\|_p^p + 1)] \end{aligned} \tag{32}$$

where Γ is a finite, N-independent constant. By combining eq. (11), (29)-(32) we deduce that we have the form

$$\lim[N-1/N] = \sum_{j=0}^{p-2} \lambda^j h_j + \frac{\text{entire function}}{\text{entire function}}\ , \tag{33}$$

where for A in C_p both entire functions in (33) satisfy

$$|\text{entire function}| \le k\exp(B\lambda^p)\ . \tag{34}$$

The modified numerator and denominator defined in this way converge separately to entire functions, and thus, of course, the Padé approximant does too. The result for the Padé approximant itself is more general than for the numerator and denominator separately as we shall see below.

COMPACT OPERATORS

First let us consider the case where P_N is an orthogonal projection ($\mathcal{S}_N = \mathcal{S}_N'$). If we subtract eq. (6) from eq. (1) we get

$$f - f_N = \lambda A (f - f_N) + \lambda (A - P_N A P_N) f_N. \qquad (35)$$

Now if $\|f_N\| < \infty$, then the second term on the right hand side of (35) goes to zero as A is compact. Then eq. (35) becomes the form

$$d \simeq \lambda A d. \qquad (36)$$

Thus, if λ is not a singular point of eq. (1), we may conclude $d = 0$ by the Fredholm alternative theorem. Hence

$$f_N \to f ,$$

$$[N-1/N] = (h, f_N) \to (h, f) = h(\lambda). \qquad (37)$$

If, on the other hand, $\|f_N\| \to \infty$ for all N, we can define

$$d_N = f_N / \|f_N\| , \qquad (38)$$

an element of unit norm. Eq. (6) becomes

$$d_N - \lambda P_N A d_N \to 0. \qquad (39)$$

As A is compact, there exists a subsequence of N's such that the limit over the subsequence exists and has the value

$$\lim \lambda A d_N = d, \tag{40}$$

which implies, as A is compact,

$$d - \lambda A d = 0. \tag{41}$$

Since we are assuming λ is not a singular point of (1), we conclude by the Fredholm alternative theorem that $\|d\| = 0$, which is a contradiction. Therefore there does not exist an infinite sequence of f_N whose norm tends to infinity, but at most a finite number of such equations. Therefore we conclude if A is a compact operator and the P_N are orthogonal projection operators that when λ is not a singular point of eq. (1)

$$\lim [N-1/N] = h(\lambda), \tag{42}$$

where the limit is taken over the infinite number of N's for which the Padé approximants exist.[3]

In the case P_N is not orthogonal, less complete results have been obtained. The problem here is that the magnitude of $\|P_N\|$ is uncontrolled. In particular, insofar as we are concerned it is only the magnitude of the element,

$$P_N A e_N, \tag{43}$$

where the e_j are defined before eq. (12), that is uncontrolled, as by construction

$$P_N A e_j \equiv A e_j, \; j \leq N - 1. \tag{44}$$

If we make the additional mild assumption that (note that this equation is misprinted in ref. 1)

$$\lim_{N\to\infty} \inf \|P_N A P_N A e_N\| = 0, \tag{45}$$

where

$$\rho_N = \sum_{j=1}^{N} e_j(e_j, \quad) \tag{46}$$

is the orthogonal projection on $\mathcal{S}_N$. Then we can prove:[1]

Theorem: Let Λ be any closed bounded region in the complex λ plane not containing a singular point of eq. (1). Then either (i) a finite order Padé approximant to $h(\lambda)$ of eq. (11) is exact, or (ii) eq. (42) holds, or (iii) for each λ in Λ for which (i) and (ii) fail, there exists an infinite subsequence of N's such that (ii) holds for all other λ in Λ.

APPLICATION TO QUANTUM SCATTERING THEORY

This application is a generalization of that of Garibotti and Villani.[7] We consider the problem of potential scattering in non-relativistic quantum mechanics. The fundamental equation is the Schrodinger equation

$$-\nabla^2\psi(\vec{r}) + \lambda V(r)\ \psi(\vec{r}) = k^2\psi(\vec{r}). \tag{47}$$

We restrict the potential function by assuming that $V(r)$ is of single sign, spherically symmetric, and satisfies

$$\int_0^\infty |V(r)|\exp(2|\nu|r)\ r dr < \infty. \tag{48}$$

Then, for the partial wave decomposition of (47) we can show: (i) a slightly recast version of the kernel of the usual corresponding integral equation is of trace class. (ii) The upper Hessenberg form of the kernel is tridiagonal. Thus the results we have reported show that the numerator and denominator of the [M/M] Padé approximants converge separately, and the denominator converges to the Jost function, as one would hope. These conclusions hold if $|Im(k)| \leq \nu$.

In as much as Fredholm equations appear very frequently throughout the field of mechanics, the potential applications of these convergence results in the area of this conference seems to me to be very large.

REFERENCES

1. G. A. Baker, Jr., J. Math. Phys. 16, 813 (1975).

2. M. Cini and S. Fubini, Nuovo Cimento 11, 142 (1954).

3. G. A. Baker, Jr., J. Math. Anal. Appl. 43, 498 (1973).

4. J. Nuttall, Phys. Rev. 157, 1312 (1967).

5. G. A. Baker, Jr., "Essentials of Padé Approximants" (Academic Press, N. Y., 1975).

6. N. Dunford and J. T. Schwartz, "Linear Operators, Part II: Spectral Theory" (Interscience, New York, 1963), Chap. 11, Sec. 9.

7. C. R. Garibotti and M. Villani, Nuovo Cimento A59, 107 (1969); A61, 747 (1969).

CONSTRUCTION OF VARIATIONAL BOUNDS FOR THE N-BODY EIGENSTATE PROBLEM BY THE METHOD OF PADE APPROXIMATIONS

D. Bessis

Service de Physique Théorique
CEN-Saclay, BP n°2 - 91190 Gif-sur-Yvette - France

We first recall how the Padé approximations applied to the resolvant of a N-body Hamiltonian generate new improved variational principles for the fundamental and the excited states, which generalize the Rayleigh-Ritz principle. In this scheme, the Rayleigh-Ritz principle is deduced from the lowest Padé Approximation, we therefore analyse completely the content of the next approximation, which gives rise to a variational principle in which is embedded the knowledge coming from more terms of the resolvant expansion.

An application to the case of N fermions interacting via a two-body potential which is itself a sum of Gaussian potentials is analysed. We show that in this case, the reconstruction of discrete eigenstates and eigenfunctions can be done in a purely algebraic way without any multiple integral calculations : the eigenstates being approximated by a monotonously decreasing sequence rapidly converging.

In the conclusion we recall how the method extends to singular potentials (hard cores) and a very simple two-body problem is tested numerically for illustration.

INTRODUCTION

In physics one is often faced with the problem of expanding a given quantity G in terms of an expansion parameter α .

$$G(\alpha) = G_o + \alpha G_1 + \ldots + \alpha^n G_n + \ldots \qquad \text{(I-1)}$$

Four cases can happen

1) The series (I-1) is, for the "physical" value of α , convergent and rapidly convergent. One can therefore consider that (I-1) solves the problem.

2) The series (I-1) is convergent but too slowly to be effective. Eq. (I-1) is therefore, as it stands, useless for this precise problem .

3) The series (I-1) is divergent for all α , but asymptotic. If the physical value of α is sufficiently small, then again (I-1) can be considered as solving the problem because the "effective convergence" could be extremely good.

4) The series (I-1) is divergent for all α , and has no asymptotic properties : (I-1) is useless as it stands.

Most of the time, one is left with cases 2 and 4, for which it is necessary to apply an "accelerator of convergence", among which the Padé Approximation is, for deep reasons, one of the most powerful.

More generally one can ask the following question :

Given a finite number G_o, G_1, ..., G_N. of coefficients in the expansion (I-1), find "the best" upper and lower bound for $G(\alpha)$,

$$\mathcal{F}_-(\alpha; G_o, G_1, \ldots, G_N) \leq G(\alpha) \leq \mathcal{F}_+(\alpha; G_o, \ldots, G_N)^*$$

How does one construct the unknown functionals $\mathcal{F}_\pm$? Of course the problem makes sense only if, not only the G_i, $0 \leq i \leq N$ are known, but also if additional properties of the function $G(\alpha)$ are given, in such a way that the bounds are achieved for $G(\alpha)$ in a precise class of functions.

For instance, if $G(\alpha)$ fulfills a dispersion relation with respect to α, without subtractions, with a positive discontinuity, that is $G(\alpha)$ is a Stieltjes function :

$$G(\alpha) = \int_o^\infty \frac{\sigma(x)\,dx}{1 + x\alpha} , \qquad \sigma(x) \geq 0 ,$$

---

*When $G(\alpha)$ is complex, one has to deal with inclusion domain see Ref. (1).

then for α real and positive the functionals $\mathcal{F}_+$ and $\mathcal{F}_-$ are given respectively[2) by the [N/N] and [N-1/N] Padé Approximants built up from the coefficients G_i $(0 \leq i \leq 2N)$ or $(0 \leq i \leq 2N-1)$.

Another example is provided by the case where

$$G(\alpha) = \int_0^\infty e^{-\alpha x} \rho(x) dx \quad , \quad \rho(x) \geq 0, \qquad (I.3)$$

namely if $G(\alpha)$ is the Laplace transform of a positive measure.

Then the functionals $\mathcal{F}_\pm$ are given by Generalized Padé Approximants [N/N] and [N-1/N] built up on the coefficients G_i [3).

An interesting fact, for those two examples, is that the functionals exist independently of the convergence or divergence of the series (I-1) which may very well be divergent for all α. The problem is best expressed in terms of information theory : given the class to which $G(\alpha)$ belongs, given a finite number of its derivatives at $\alpha = 0$, what is the best upper and lower bound which one can expect for $G(\alpha)$.

Even more interesting is the case of variational bounds, that is the case where, for some reasons, the G_i themselves do not exist and need regularization, for instance in the so-called case of singular interactions[4). Then if $G(\alpha)$ is regularized into $G^\varepsilon(\alpha)$ by means of a small regularization parameter ε and if furthermore :

$$G(\alpha) \leq G^\varepsilon(\alpha) \quad , \qquad (I-4)$$

if finally we can derive the functional $\mathcal{F}_+^\varepsilon(\alpha ; G_o^\varepsilon ; G_1^\varepsilon ; \ldots, G_N^\varepsilon)$ for $G^\varepsilon(\alpha)$, then we find

$$G(\alpha) \leq G^\varepsilon(\alpha) \leq \mathcal{F}_+^\varepsilon(\alpha ; G_o^\varepsilon ; G_1^\varepsilon ; \ldots, G_N^\varepsilon). \qquad (I-5)$$

This gives the variational bound :

$$G(\alpha) \leq \inf_\varepsilon \mathcal{F}_+^\varepsilon (\alpha ; G_o^\varepsilon ; G_1^\varepsilon ; \ldots, G_N^\varepsilon) \qquad (I-6)$$

Even though the regularized Taylor coefficients of $G(\alpha)$ tend to infinity when $\varepsilon \to 0$, the functional $\mathcal{F}_+^\varepsilon$ will have a minimum for some $\bar{\varepsilon}_N$, which at that value of N will depend on α. This $\bar{\varepsilon}_N(\alpha)$ is the best choice of the cut-off regulator for the information contained in the Taylor series stopped at order N. Such a scheme has been extensively used in ref. (4).

Let us come now to a very well known case. Suppose we consider the mean value of the resolvant of a Hermitian operator H in Hilbert space, having a negative discrete spectrum E_o, $E_1, \ldots$, and a positive continuous spectrum, typically a N-body Hamiltonian in Quantum Mechanics. Then the resolvant reads :

$$\mathcal{R}_\varphi(z) = \langle\varphi| \frac{z}{1+zH} |\varphi\rangle \qquad \text{(I-7)}$$

and if we expand it formally, we get :

$$\mathcal{R}_\varphi(z) = z \sum_{n=0}^{\infty} (-z)^n \langle\varphi|H^n|\varphi\rangle \qquad \text{(I-8)}$$

The operator H being unbounded in general, the moments $\langle\varphi|H^n|\varphi\rangle$ are all infinite, except if $|\varphi\rangle$ belongs to the domain of all the powers of H. Let us suppose first that we choose $|\varphi\rangle$ in such a way that $\langle\varphi|H|\varphi\rangle$ exists.

Then, the Rayleigh-Ritz variational principle asserts that the ground state energy E_o fulfills :

$$E_o \leq \frac{\langle\varphi|H|\varphi\rangle}{\langle\varphi|\varphi\rangle} \qquad \forall \; |\varphi\rangle \in \mathcal{H} \;, \qquad \text{(I-9)}$$

and $E_o = \inf\limits_{|\varphi\rangle} \dfrac{\langle\varphi|H|\varphi\rangle}{\langle\varphi|\varphi\rangle}$; (I-10)

Therefore the knowledge of only the <u>first</u> term of the expansion (I-8) allows us to build up a variational functional for the lowest eigenstate of H. It is important to remark that the series (I-8) has a <u>zero</u> radius of convergence for a generic $|\varphi\rangle$: this is so because $R_\varphi(z)$ is singular at z equal zero due to the fact that the spectrum of H extends to $+\infty$. The point $z = 0$ being a singularity of $R_\varphi(z)$ the radius of convergence of (I-8) is of course zero. Nevertheless the bound (I-9) holds.

Now the interesting question is :

Given more moments : $\mu_R = \langle\varphi|H^k|\varphi\rangle$ (I-11)

of the expansion (I-8), can one improve the bound (I-9) and more generally can one derive variational principles not only for the ground state E_o, but also for E_1, $E_2, \ldots$ the so-called excited states ? The answer is yes, and it is the Padé Approximation technique which is the generating tool of these variational improved Rayleigh-Ritz principles : we shall call them the Padé- Rayleigh-Ritz principles. Furthermore, these bounds can be shown to be the best possible[5)] which, one can construct from the knowledge of only a finite number of moments μ_k :

II - <u>THE PADE-RAYLEIGH-RITZ PRINCIPLES.</u>

Given an <u>even</u> number of moments $\mu_o, \mu_1, \ldots, \mu_{2N-1}$.

We construct the following determinant (which is the denominator of the [N/N] Padé approximation built up on the resolvant)

$$\Delta_{\varphi}(E,N) = \begin{vmatrix} \mu_0 & \mu_1 & \cdots & \mu_N \\ \mu_1 & \mu_2 & \cdots & \mu_{N+1} \\ \cdots & \cdots & \cdots & \cdots \\ \mu_{N-1} & \mu_N & \cdots & \mu_{2N-1} \\ 1 & E & \cdots & E^N \end{vmatrix} \qquad \text{(II-1)}$$

$\Delta_{\varphi}(E,N)$ is clearly a polynomial of degree N in E. It can be shown[5] that the N roots of $\Delta_{\varphi}(E,N)$ are all real and distinct and therefore we can order them :

$$E_0^{(N)} < E_1^{(N)} < \ldots < E_p^{(N)} < \ldots < E_{N-1}^{(N)} \quad . \qquad \text{(II-2)}$$

Furthermore $E_p^{(N)}$ is an upper bound to the pth excited state of H[5].

$$\begin{cases} E_0 \leq E_0^{(N)} \\ E_1 \leq E_1^{(N)} \\ E_i \leq E_i^{(N)} \\ \cdots\cdots\cdots \\ E_N \leq E_{N-1}^{(N)} \end{cases} \qquad \text{(II-3)}$$

The true eigenvalues E_p are of course independent of the test vector $|\varphi\rangle$ we choose, but clearly the bound $E_p^{(N)}$ is dependent on $|\varphi\rangle$ because the moments μ_k are dependent on $|\varphi\rangle$. Therefore

$$E_L \leq \inf_{|\varphi\rangle} E_L^{(N)}(\varphi) \quad . \qquad \text{(II-4)}$$

More precisely we have[4]

$$E_L = \inf_{|\varphi\rangle} E_L^{(N)}(\varphi) \qquad \text{(II-5)}$$

In that way, we generate variational principles for the excited states. A further interesting fact is that if we now increase N by one unit, one proves that[5] :

$$E_L^{(N+1)}(\varphi) \leq E_L^{(N)}(\varphi) . \qquad \text{(II-6)}$$

In other words, by adding more moments, we are sure to improve the variational principle for the ith excited state.

$$E_i \leq \ldots \leq E_i^{(N+1)}(\varphi) \leq E_i^{(N)}(\varphi) \leq \ldots \leq E_i^{(i+1)}(\varphi) . \qquad \text{(II-7)}$$

Even if we do not use the variational properties in $|\varphi\rangle$ of the $E_L^{(N)}(\varphi)$ bounds, and consider the sequence of the $E_L^{(N)}(\varphi)$ <u>at fixed</u> $|\varphi\rangle$, the sequence decreases monotonously towards E_L and in cases of practical interest ($|\mu_k| \lesssim (2k)!$) the sequence converges very rapidly towards E_L.

However here, we do not want to make use of the convergence of the bounds (II-7), (because for the N-body problem, for instance, it is not easy to compute too high moments) but rather use them as new improved variational principles.

For N=1, we have

$$\Delta_\varphi(E,1) = \begin{vmatrix} \mu_o & \mu_1 \\ 1 & E \end{vmatrix} = \mu_o E - \mu_1 \qquad \text{(II-8)}$$

Then

$$E_o^{(1)} = \frac{\mu_1}{\mu_o} = \frac{\langle\varphi|H|\varphi\rangle}{\langle\varphi|\varphi\rangle} \qquad \text{(II-9)}$$

That is using (II-5)

$$E_o = \inf_{|\varphi\rangle} E_o^{(1)} = \inf_{|\varphi\rangle} \frac{\langle\varphi|H|\varphi\rangle}{\langle\varphi|\varphi\rangle} \qquad \text{(II-10)}$$

we find the Rayleigh-Ritz variational principle.

III - <u>THE CASE N=2, OR THE PADE-RAYLEIGH-RITZ VARIATIONAL PRINCIPLE</u>.

For N=2 we get

$$\Delta_\varphi(E,2) = \begin{vmatrix} \mu_o & \mu_1 & \mu_2 \\ \mu_1 & \mu_2 & \mu_3 \\ 1 & E & E^2 \end{vmatrix} \quad ; \quad \mu_k = \langle\varphi|H^k|\varphi\rangle \qquad \text{(III-1)}$$

Introducing the deviations :

$$\delta_2 = \frac{\langle\varphi|[H-\bar{H}]^2|\varphi\rangle}{\langle\varphi|\varphi\rangle} \qquad \bar{H} = \frac{\mu_1}{\mu_o} = \frac{\langle\varphi|H|\varphi\rangle}{\langle\varphi|\varphi\rangle} \qquad \text{(III-2)}$$

$$\delta_3 = \frac{\langle\varphi|[H-\bar{H}]^3|\varphi\rangle}{\langle\varphi|\varphi\rangle} \qquad \text{(III-3)}$$

we get, solving $\Delta_\varphi(E,2) = 0$

$$\begin{cases} E_o^{(2)} = \dfrac{\langle\varphi|H|\varphi\rangle}{\langle\varphi|\varphi\rangle} - \left\{ \sqrt{\left(\dfrac{\delta_3}{2\delta_2}\right)^2 + \delta_2} - \dfrac{\delta_3}{2\delta_2} \right\} \\ E_1^{(2)} = \dfrac{\langle\varphi|H|\varphi\rangle}{\langle\varphi|\varphi\rangle} + \sqrt{\left(\dfrac{\delta_3}{2\delta_2}\right)^2 + \delta_2} + \dfrac{\delta_3}{2\delta_2} \end{cases} \qquad \text{(III-4)}$$

In particular we obtain for the ground state E_o the improved variational principle :

$$E_o = \inf_{|\varphi\rangle} \left\{ \frac{\langle\varphi|H|\varphi\rangle}{\langle\varphi|\varphi\rangle} - \left[\sqrt{(\frac{\delta_3}{2\delta_2})^2 + \delta_2} - \frac{\delta_3}{2\delta_2}\right]\right\} \qquad \text{(III-5)}$$

we remark that, δ_2 being always positive, the quantity :

$$\varepsilon_2(\varphi) = \left[\sqrt{(\frac{\delta_3}{2\delta_2})^2 + \delta_2} - \frac{\delta_3}{2\delta_2}\right] \qquad \text{(III-6)}$$

is always positive, and therefore comes <u>subtractively</u> to the Rayleigh-Ritz ordinary principle : it can be considered as the correction to it, induced by the extra information coming from the knowledge of the second and third moment.

By changing the sign of the square root, we get the variational principle for the first excited state :

$$E_1 = \inf_{|\varphi\rangle} \left\{ \frac{\langle\varphi|H|\varphi\rangle}{\langle\varphi|\varphi\rangle} + \sqrt{(\frac{\delta_3}{2\delta_2})^2 + \delta_2} + \frac{\delta_3}{2\delta_2} \right\} \qquad \text{(III-7)}$$

On (III-5), we see that if δ_2 is small,that is if $|\varphi\rangle$ is already near to the exact ground state eigenfunction , the correction $\varepsilon_2(\varphi)$ is small; if on the contrary δ_2 is very large, the correction $\varepsilon_2(\varphi)$, being an increasing function of δ_2, becomes large and that large correction will improve the bound significantly. Even if δ_2 is small, the precision on E_o is very much improved. As shown on numerical examples, factors as large as 10^4 on the precision are gained by using the Padé principle with respect to the usual Ritz principle.

Finally we want to point out the important fact that $\varepsilon_2(\varphi)$ is also a monotonously increasing function of δ_3, as checked easily. Therefore if we replace δ_3 by a rough upper bound, (or equivalently μ_3 by an upperbound), we still get a variational principle for the ground state :

$$E_o = \inf_{|\varphi\rangle} \left\{ \frac{\langle\varphi|H|\varphi\rangle}{\langle\varphi|\varphi\rangle} - \left[\sqrt{(\frac{\delta_3^+}{2\delta_2})^2 + \delta_2} - \frac{\delta_3^+}{2\delta_2}\right]\right\} \qquad \text{(III-8)}$$

where δ_3^+ is an upper bound of δ_3 , obtained by using an upper bound for μ_3. Eq. (III-8) shows also clearly that, if only δ_2 is known, then it is not possible to improve the Ritz principle, because when $\delta_3 \to +\infty$, $\varepsilon_2(\varphi) \to 0$. Therefore the knowledge of δ_3, even through a very crude estimate, is fundamental in this method. (see a physical consequence of this last statement in section V).

IV - THE EIGEN-FUNCTION APPROXIMATION.

Let us suppose, from now on, that we have normalized the test vector $|\varphi\rangle$:

$$\langle\varphi|\varphi\rangle = +1. \qquad \text{(IV-1)}$$

Then, following ref. (5), the approximate ground state eigenfunction, for the N=2 case is :

$$|\Phi_o^{(2)}\rangle = |\varphi\rangle - \frac{\varepsilon_2}{\delta_2}(H-\bar{H})|\varphi\rangle \qquad \text{(IV-2)}$$

It is easy to check that, if we use $|\Phi_o^{(2)}\rangle$ as a test vector in the Rayleigh-Ritz variational principle, we get the Padé-Ritz principle that is :

$$\frac{\langle\Phi_o^{(2)}|H|\Phi_o^{(2)}\rangle}{\langle\Phi_o^{(2)}|\Phi_o^{(2)}\rangle} = \langle\varphi|H|\varphi\rangle - \left[\sqrt{\left(\frac{\delta_3}{2\delta_2}\right)^2 + \delta_2} - \frac{\delta_3}{2\delta_2}\right] \qquad \text{(IV-3)}$$

The reader will notice that $|\Phi_o^{(2)}\rangle$ and $E_o^{(2)}$ are the ground state eigenvalue and eigenvector of $P_2\, H\, P_2$, where P_2 is the projector onto the two-dimensional space $\mathscr{E}^{(2)}$ spanned by the vectors $|\varphi\rangle$ and $H|\varphi\rangle$. This remark can be completely generalized, and is linked to the fact that the Lanczos method for matrices, and its generalization to Hilbert space the tri-diagonalisation Jacobi method, are deeply linked with the theory of Padé Approximation[6].

To end up this section we give the formulae generalizing (IV-2) for the Nth approximation.

Introducing the normalized polynomials :

$$\bar{\Delta}(E,N) = \frac{\Delta(E,N)}{\sqrt{G_N G_{N-1}}} \quad ; \quad G_N = \begin{vmatrix} \mu_o \cdots\cdots \mu_N \\ \vdots \\ \mu_N \cdots\cdots \mu_{2N} \end{vmatrix} > 0. \qquad \text{(IV-4)}$$

The Lth approximated eigenfunction at order N is given by :

$$|\Phi_L^{(N)}\rangle = \sum_{j=0}^{N-1} \bar{\Delta}(E_i^{(N)}, j)\; \bar{\Delta}(H, j)\, |\varphi\rangle \ . \qquad \text{(IV-5)}$$

The $|\Phi_L^N\rangle$ are obtained by diagonalizing the Hermitian matrix

$$H_N = P_N\, H\, P_N \qquad \text{(IV-6)}$$

where P_N is the projector onto the N-dimensional space spanned by the vectors $|\varphi\rangle$, $H|\varphi\rangle, \ldots H^{N-1}|\varphi\rangle$.

While the $|\Phi_L^{(N)}\rangle$ are orthogonal but not normalized, the $\bar{\Delta}(H,j)|\varphi\rangle$ are a set of orthonormal vectors because it is known that the $\bar{\Delta}(E,N)$ form a set of orthogonalized polynomials with respect to the spectral measure of the operator H[5].

V - APPLICATION TO A SYSTEM OF N FERMIONS.

To apply effectively the Padé-Ritz principle (IV-5) it is necessary to find a way to compute easily the moments

$$\mu_k = \langle\varphi|\, H^k\, |\varphi\rangle^{*}. \qquad \text{(V-1)}$$

In the Z-body problem the moments are given Z-uple integrals which are difficult or even impossible to compute if k is large. However, there are cases in which the moments can be reduced to entirely algebraic expressions.

An example is given in ref. (5), for the case of the most general d-dimensional anharmonic oscillator, by choosing a suitable variational test vector $|\varphi\rangle$.

Here we want to consider the case of Z fermions interacting via a two body potential :

$$\tilde{H}_Z = \sum_{i=1}^{Z} \frac{\vec{p}_2^{\,2}}{2m_i} + \sum_{i<j\leq Z} V(|\vec{r}_i-\vec{r}_j|) - \frac{(\sum_{i=1}^{Z}\vec{p}_i)^2}{2M} \qquad \text{(V-2)}$$

where we have subtracted the center of mass energy.

We shall suppose that V, the two-body potential is a superposition of Gaussian potentials :

$$V(|\vec{r}_i-\vec{r}_j|) = \sum_{\rho=1}^{\rho=S} \bar{P}_\rho([\vec{r}_i-\vec{r}_j]^2)\; e^{-\mu_\rho(\vec{r}_i-\vec{r}_j)^2} \qquad \text{(V-3)}$$

(where $\bar{P}_\rho(x)$ is a polynomial in x). With such a superposition, we can build up a large class of phenomenological potentials. The interesting case of hard cores will be postponed to a further section.

We must now choose the test vector in such a way that

1) The Z-uple integral reduces to an algebraic calculation.

2) The anti-symetrization principle (Pauli-principle) is automatically fulfilled.

---

* $|\varphi\rangle$ is a given vector, by using "matrix" Padé Approximant most of the results exposed here generalize to the case where $|\varphi\rangle$ is a set of vectors $|\varphi_1\rangle$, $|\varphi_2\rangle$ ... $|\varphi_R\rangle$, generating a "model" space.

We shall choose $|\varphi\rangle$ to be an arbitrary sum of Slater determinants built up with the states of the harmonic oscillator.

$$\langle\vec{r}_1,\vec{r}_2,\ldots,\vec{r}_Z\rangle = \sum_{t=1}^{P} \omega_{\vec{t}} \frac{1}{\sqrt{Z!}} \begin{vmatrix} \varphi_o(\vec{r}_1,\vec{t})\ldots\varphi_o(\vec{r}_Z,\vec{t}) \\ \varphi_1(\vec{r}_1,\vec{t})\ldots\varphi_1(\vec{r}_Z,\vec{t}) \\ \cdots\cdots\cdots\cdots\cdots \\ \varphi_{Z-1}(\vec{r}_1,\vec{t})\ldots\varphi_{Z-1}(\vec{r}_Z-\vec{t}) \end{vmatrix} \tag{V-4}$$

where the $\varphi_m(\vec{r},\vec{t})$ are a set of orthonormalized harmonic oscillator eigenfunctions :

$$\varphi_m(\vec{r},\vec{t}) = \varphi_{m_1}(x,t_1).\ \varphi_{m_2}(y,t_2)\ .\ \varphi_{m_3}(z,t_3) \tag{V-5}$$

with

$$\varphi_{m_1}(x,t_1) = \left[\sqrt{\pi}\ 2^m\ m!\ t_1^{-1}\right]^{-1/2} H_{m_1}(t_1x)\ e^{-\frac{t_1^2}{2}x^2} \tag{V-6}$$

and analogous expressions for $\varphi_{m_2}(y,t_2)$, $\varphi_{m_3}(z,t_3)$. The $H_m(\rho)$ are the usual Hermite polynomials.

The parameters $\omega_{\vec{t}}$ and $\vec{t}$ are the variational parameters to be used in the calculation.

Combining all these informations and taking into account the fact that the derivative of a polynomial times a Gaussian is again a polynomial times a Gaussian we see that the moment μ_k will be of the form :

$$\mu_k = \sum_{\alpha} \int d\vec{r}_1\ldots d\vec{r}_Z\ P_\alpha(\vec{r}_1,\ldots\vec{r}_Z)\ e^{-Q_\alpha(\vec{r}_1,\ldots,\vec{r}_Z)} \tag{V-7}$$

where Q_α is a quadratic <u>positive</u> form of the $\vec{r}_i$:

$$Q_\alpha = \sum_{i=j=1}^{Z} q_{ij}^{\alpha}\ \vec{r}_i.\vec{r}_j. \tag{V-8}$$

and P_α is a polynomial in the variables $\vec{r}_1,\ldots,\vec{r}_Z$.

Therefore, we are left with integrals of the form :

$$\int_{-\infty}^{+\infty} dx_1\ldots dx_Z\ (x_{i_1}x_{j_1})^{\alpha_1}(x_{i_2}x_{j_2})^{\alpha_2}\ldots(x_{i_m}x_{j_m})^{\alpha_m} \exp\left\{-\sum_{i,j=1}^{Z} q_{ij}x_ix_j\right\}$$

(V-9)

which are equal to

$$(-)^{\alpha_1+\alpha_2\dots+\alpha_m} \frac{\partial^{\alpha_1+\alpha_2+\dots+\alpha_m}}{\partial^{\alpha_1} q_{i_1j_1}\dots\partial^{\alpha_m} q_{i_mj_m}} \left\{ \frac{\pi^{Z/2}}{(\mathrm{Det}\, q_{ij})^{1/2}} \right\} \quad \text{(V-10)}$$

In principle the calculation is straighforward and purely algebraic. In practice the enormous number of terms to consider limits the possibilities to the calculation of the first three moments. Even for the third moment one may have to consider one thousand terms contributing. This means that it is necessary to use the variational Padé-Ritz principle with the upper bound on δ_3. As a consequence it is important to know how to get not too bad upper bounds for

$$\delta_3 = \langle\varphi\, [(T-\bar{T}) + (V-\bar{V})]^3 |\varphi\rangle. \quad \text{(V-11)}$$

This problem will be considered in future work.

VI - <u>THE CASE OF AN INFINITE NUMBER OF BODIES</u>

Let us consider the case where the number Z of bodies tends to infinity. Let us suppose that the ground state energy per body, $E_o^{(Z)}/Z$ has a finite limit when Z tends to infinity and that this limit E_o is such that

$$E_o \leq \frac{E_o^{(Z)}}{Z} \qquad \text{for} \quad Z > Z_o. \quad \text{(VI-1)}$$

This case seems to be not unlikely for nuclear physics. Then we can use the Rayleigh-Ritz principle and write :

$$E_o \leq \frac{E_o^{(Z)}}{Z} \leq \frac{\langle\varphi|H_Z|\varphi\rangle}{Z\langle\varphi|\varphi\rangle} \quad \text{(VI-2)}$$

or

$$E_o = \inf_Z \inf_{|\varphi\rangle} \frac{\langle\varphi|H_Z|\varphi\rangle}{Z\langle\varphi|\varphi\rangle} \quad \text{(VI-3)}$$

If we use the Padé-Ritz principle we get

$$E_o = \inf_Z \inf_{|\varphi\rangle} \left[\frac{\langle\varphi|H_Z|\varphi\rangle}{Z\langle\varphi|\varphi\rangle} - \frac{1}{Z}\left\{ \sqrt{\left[\frac{\delta_3^+(Z)}{2\delta_2(Z)}\right]^2 + \delta_2(Z)} - \frac{\delta_3^+(Z)}{2\delta_2(Z)} \right\}\right]$$

It is interesting to notice that while we can replace (VI-3) by :

$$E_o = \inf_{|\varphi\rangle} \lim_{Z\to\infty} \frac{\langle\varphi|H_Z|\varphi\rangle}{\langle\varphi|\varphi\rangle} \qquad \text{(VI-5)}$$

because we expect the limit for not too bad $|\varphi\rangle$ to exist and be meaningfull, in general the limit of the correction term when $Z \to \infty$ will be zero[7], and so no improvement will be achieved by the Padé-Ritz principle if we take brutally the limit $Z\to\infty$. Instead, if we look for the inf in Z, it is clear that this inf will give a better upper bound to E_o than the Rayleigh-Ritz principle, as best shown on Fig. 1.

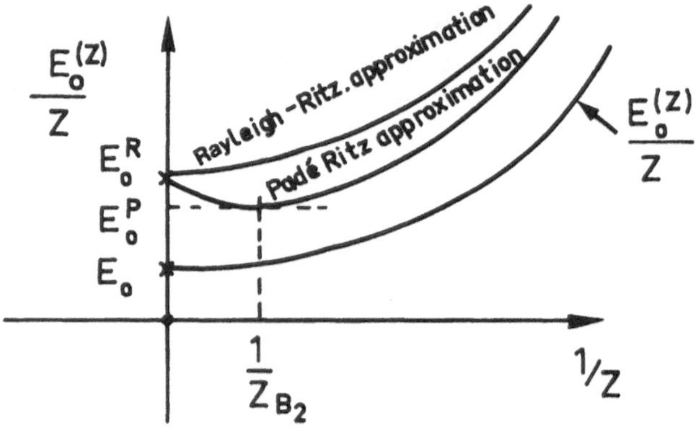

Fig.1 - $E_o^{(Z)}/Z$ as a function of 1/Z. and its approximations.

We see that $E_{o,P}$ (the Padé-Ritz approximation) will be <u>always</u> better than $E_{o,R}$ the Rayleigh-Ritz Approximation .

It will be for some value Z_{B_2} (Z best) that we achieve the best upper bound for E_o. This technique clearly uses the number of bodies itself as variational parameter. If one takes the next Padé Approximation, one gets a new Z_{B_3} which gives an improved upper bound for E_o. It is possible to proove that this sequence of minimum converges to E_o in very general cases[4].

A very analogous technique can be used to treat the case of hard core potentials : one regularizes the potential and then uses the cut-off as a variational parameter, see ref. (4) and (5) for a detailed discussion.and forthcoming paper.

Before ending this section we want to point out the important fact, that, as explained at the end of section III the knowledge of only δ_2 cannot improve the calculation by no means, and that some knowledge of δ_3 (upper bound) is necessary. But up to third order, the deviations (III-2), (III-3) are identical to the cumulants which occur in the Ursell-Mayer cluster expansion. This suggests that (in that scheme) some knowledge of the three-body correlations is necessary, together with that of two-body correlations (Brueckner) to improve, in a variational framework, the result one can obtain from an independent particles approximation.

VI - A NUMERICAL EXAMPLE.

As numerical example, let us take the one dimensional harmonic oscillator :

$$H = p^2 + x^2 \quad ; \quad p = -i \frac{d}{dx} \tag{VI-1}$$

The ground state of which is known exactly to be

$$E_o = 1 \tag{VI-2}$$

As test vector we take

$$\langle x|\varphi\rangle = e^{-\frac{\alpha}{2}(x-\beta)^2} \tag{VI-3}$$

The exact eigenfunction is obtained for $\alpha=1$ and $\beta=0$.

The deviations are easily computed for (VI-3) and one gets :

$$\bar{H} = \frac{\langle\varphi|H|\varphi\rangle}{\langle\varphi|\varphi\rangle} = \frac{1}{2}(\alpha + \frac{1}{\alpha}) + \beta^2 . \tag{VI-4}$$

$$\delta_2 = \frac{1}{2}(\alpha - \frac{1}{\alpha})^2 + \frac{2\beta^2}{\alpha} \tag{VI-5}$$

$$\delta_3 = (\alpha + \frac{1}{\alpha})(\alpha - \frac{1}{\alpha})^2 + 2\beta^2(\frac{3}{\alpha^2} - 1) \tag{VI-6}$$

We shall take α as a variational parameter and β as a fixed parameter.

Then we get :

$$E_{\text{Rayleigh-Ritz}} = \inf_{0\leq\alpha} \frac{1}{2}(\alpha + \frac{1}{\alpha}) + \beta^2 \tag{VI-7}$$

$$E_{\text{Padé-Ritz}} = \inf_{0\leq\alpha} \left\{ \frac{1}{2}(\alpha+\frac{1}{\alpha}) + \beta^2 - \left[\sqrt{(\frac{\delta_3}{2\delta_2})^2 + \delta_2} - \frac{\delta_3}{2\delta_2}\right]\right\} \tag{VI-8}$$

For $\beta = 0$ we have :

$$\left.\begin{aligned} E_{RR} &= \frac{1}{2}(\alpha + \frac{1}{\alpha}) \\ E_{PR} &= \frac{3}{2}(\alpha + \frac{1}{\alpha}) - \sqrt{\frac{3}{2}(\alpha + \frac{1}{\alpha})^2 - 2} \end{aligned}\right\} \tag{VI-9}$$

We see that while only the first derivative of $E_{RR}(\alpha)$ vanishes at $\alpha = 1$, the first two derivatives of $E_{RR}(\alpha)$ vanish at $\alpha = 1$. Of course in this case both minima are equal and equal to the exact value.

We give now a little summary of the numerical results for three typical values of β .

$2\beta^2$	Exact value	Rayleigh-Ritz value	Padé-Ritz value	Improvement factor.
0	1	1. (exact)	1. (exact)	$\frac{0}{0} = \infty$
0.01	1	1.005	1.00000025	20 000
0.1	1	1.05	1.000195	250
1.	1	1.5	1.0583	8

We have defined the improvement factor by :

$$I = \frac{E_{Rayleigh-Ritz} - E_{exact}}{E_{Padé-Ritz} - E_{exact}} \simeq \frac{2}{\beta_2} \quad \text{for } \beta \text{ small.} \qquad \text{(VI-10)}$$

We see, that, even when the Rayleigh-Ritz method becomes meaningless for $2\beta^2 = 1$, the Padé-Ritz method still gives no more than a 6% error.

CONCLUSION

The method presented here has the great advantage to be variational and therefore completely independent of how big would be the coupling constant in the Hamiltonian

$$H = H_o + \lambda V$$

for a N-body system. It can be thought as representing what would be "the next corrections" to the Rayleigh-Ritz variational principle. It necessitates to build up those new improved corrected variational principles, the knowledge of more perturbative terms in the formal expansion of the resolvant of H.

When we deal with N-fermions and a two body potential superposition of Gaussian potentials we have seen that all calculations can be done algebraically and a simple numerical easily testable example as shown that improving factor as large as two to four order of magnitude in the precision can be expected.

The problem when dealing with hard-core potentials can be looked in Ref. 4 : it will be fully analyzed in a forthcoming paper.

ACKNOWLEDGMENTS

The author would like to thank Prof. B. Giraud, R. Schaeffer and A. Zuker for fruitful comments and discussions.

Dr. G. Vichniac is thanked for his help in working out completely the numerical example.

REFERENCES.

(1) - G. Baker Jr., in "Essentials of Padé Approximants". Academic Press New York 1975. Chapter 17.

(2) - See Ref. (1) Chapter 15.

(3) - J.C. Wheeler and R.G. Gordon, Bounds for averages using moment constraint in "The Padé Approximant in Theoretical Physics". Baker Jr. and Gammel editors, Mathematics in Science and Engineering Academic Press 1970.

(4) - D. Bessis, L. Epele and M. Villani, J. Math. Phys. 15, p. 2071, (1974).

(5) - D. Bessis and M. Villani, J. Math. Phys. 16, p. 462 (1975).

See also Michael Barnsley, The bounding properties of the multipoint Padé approximant to a series of Stieltjes, J.M.P. Vol. 14 n°3 March 1975.

(6) - See Appendix H of ref. (5).

(7) - D. Bessis, P. Moussa and M. Villani, Monotonous converging variational approximations to functional integrals in Quantum Statistical Mechanics. Saclay preprint D.Ph.T/75-6. January 1975 to be published in Journal of Mathematical Physics.

RATIONAL POLYNOMIAL APPROXIMANTS
IN N VARIABLES

by

J. S. R. Chisholm,
Mathematical Institute,
The University of Kent at Canterbury,
Canterbury, Kent, England.

I am reviewing work on many-variable approximants carried out by a group of applied mathematicians at the University of Kent, Canterbury, since September, 1972. I reviewed a considerable part of this work in a talk given in Marseille[1] in 1973, and much of it is now published. In writing this account, I shall therefore give only those technical details which are necessary for a broad understanding of the approximation method, and shall try to put into perspective the work that we have done; this review is necessarily a view based to some extent on my personal recollections. It is important to realise that, after the production of my initial paper[2], a group of about six were working and discussing problems almost every day throughout the winter of 1972-3. There was, in consequence, a considerable amount of give-and-take in the development of work, although the authorship of the various papers gives a very fair idea of the contribution made by each member of the group. We also benefited from some discussions with other members of the School of Mathematics.

During my visit to Texas in 1965-6, John Gammel and I several times discussed the possibility of inventing a two-variable generalisation of Padé Approximants, and I looked at the problem on a number of occasions between 1966 and 1972. In 1972, John Gammel, with Charles Critchfield, proposed a two-variable generalisation[3], but noted that their scheme differed in certain basic respects from the Padé method for a single variable. A few weeks later, I made another attempt to solve the problem.

The core of the problem seemed to me to be the definition of an analogue of *diagonal* Padé approximants, since these approximants possessed several characteristic properties which should be shared by any two-variable generalisation, and were therefore the most closely restricted class. There were other properties of two-variable approximants which also seemed very desirable, for which there was no analogue for a single variable. The properties I wished to satisfy were:

(i) Defining Property

Using the notation $\underset{\sim}{z} = (z_1,z_2)$, $\underset{\sim}{\gamma} = (\gamma_1,\gamma_2)$ and $\underset{\sim}{\infty} = (\infty,\infty)$, a double power series

$$f(\underset{\sim}{z}) = \sum_{\underset{\sim}{\gamma}=\underset{\sim}{0}}^{\underset{\sim}{\infty}} c_{\underset{\sim}{\gamma}} \, z^{\underset{\sim}{\gamma}}$$

is given. The diagonal approximant is to be a rational polynomial function

$$f_{m/m}(z) = \frac{\sum_{\underset{\sim}{\alpha}\subset P_m} a_{\underset{\sim}{\alpha}} \, \underset{\sim}{z}^{\alpha}}{\sum_{\underset{\sim}{\alpha}\subset P_m} b_{\underset{\sim}{\beta}} \, z^{\underset{\sim}{\beta}}}, \tag{2}$$

where P_m is a finite set of lattice points with $\alpha_r (r=1,2)$ taking values on set $\{0,1,2,\ldots\}$. The ratios of the coefficients $a_{\underset{\sim}{\alpha}}$ and $b_{\underset{\sim}{\beta}}$ are to be determined by a set of linear equations, formed by equating to zero the coefficients of $\underset{\sim}{z}^{\underset{\sim}{\lambda}}$ in the expression

$$E(\underset{\sim}{z}) = \left[\sum_{P_m} b_{\underset{\sim}{\beta}} z^{\underset{\sim}{\beta}}\right]\left[\sum_{\underset{\sim}{\gamma}=\underset{\sim}{0}}^{\underset{\sim}{\infty}} c_{\underset{\sim}{\gamma}} z^{\underset{\sim}{\gamma}}\right] - \sum_{P_m} a_{\underset{\sim}{\alpha}} z^{\underset{\sim}{\alpha}}, \tag{3}$$

$\underset{\sim}{\lambda}$ taking values corresponding to all lattice points in some set Q_m.

(ii) Symmetry Property

In order to preserve symmetry between z_1 and z_2 in the definitions, the sets P_m and Q_m should be symmetric between the pairs of suffixes, for example, between λ_1 and λ_2.

(iii) Existence and Uniqueness

The number of points in the set Q_m should be twice the number in P_m, less one, to provide the correct number of equations. The determinant of coefficients should, in general, be non-zero.

(iv) Homographic Covariance

Suppose that the substitutions

$$z_r = \frac{A_r w_r}{1+B_r w_r}, \qquad (r=1,2) \tag{4}$$

where A_r are any non-zero complex numbers and B_r are any complex numbers, are made in the series (1), and formal term-by-term expansions give a series

$$g(\underset{\sim}{w}) = \sum_{\underset{\sim}{\gamma}=\underset{\sim}{0}}^{\infty} d_{\underset{\sim}{\gamma}} \underset{\sim}{w}^{\underset{\sim}{\gamma}} = \sum_{\underset{\sim}{\gamma}=\underset{\sim}{0}}^{\infty} c_{\underset{\sim}{\gamma}} \prod_{r=1}^{2} \left(\frac{A_r w_r}{1+B_r w_r} \right)^{\gamma_r} . \quad (5)$$

Then if $g_{m/m}(\underset{\sim}{w})$ is the approximant defined from the series $g(\underset{\sim}{w})$, it should be given by

$$g_{m/m}(\underset{\sim}{w}) = f_{m/m}\left(\frac{A_1 w_1}{1+B_1 w_1}, \frac{A_2 w_2}{1+B_2 w_2} \right) . \quad (6)$$

<u>(v) Reciprocal Covariance</u>

The [m/m] approximant formed from the reciprocal of the series (1) should equal

$$[f_{m/m}(z)]^{-1} .$$

<u>(vi) Projection Property</u>

If z_2=0, the series (1) becomes a series in the single variable z_1. The Padé approximant to this reduced series should equal $f_{m/m}(z_1,0)$.

<u>(vii) Convergence in Measure</u>

There should be an analogue of Nuttall's theorem[4] on convergence in measure of diagonal sequences of Padé approximants.

It was not possible to consider convergence theorems before defining the approximants, so the problem was one of choosing the lattice regions P_m and Q_m so that conditions (i)-(vi) were satisfied. The problem was like solving a jig-saw puzzle : difficult and incomprehensible while one examined one or two pieces at a time, but as the pieces began to fit together, leading rapidly to a solution; apart from a relatively minor problem over symmetry, the definition of two-variable diagonal approximants satisfying (i)-(vi) emerged after two or three days work, and was then submitted for publication. At this stage, though, existence and uniqueness of the approximants had not been properly studied.

The region P_m that I eventually chose was the square $0 \leqslant \lambda_r \leqslant m$, denoted by S_1 in Figure 1. The number of points in Q_m had to be $2(m+1)^2-1$ to satisfy property (iii); the projection property (vi) could be satisfied if Q_m contained the points up to (2m,0) and (0,2m) along the "axes" of the lattice. These two requirements suggested that Q_m might be a triangular-shaped region such as $S_1+S_2+S_3$ in Figure 1. This left me short of m equations, however; the points (2m+1,0) and (0,2m+1) could not be used, since they would usually lead to overdetermination of the Padé approximants to $f(z_1,0)$ and $f(0,z_2)$; excluding these points, the next line S_4, with equation

$$\lambda_1 + \lambda_2 = 2m + 1 ,$$

contained 2m lattice points, out of which I had to select m. To preserve symmetry, I thought of equating to zero the *sum* of coefficients in $E(\underset{\sim}{z})$ corresponding to pairs of points

$$(\lambda, 2m+1-\lambda) \quad , \quad (2m+1-\lambda, \lambda) \tag{7}$$

with $\lambda = m+1,\ldots,2m$; this would provide m "symmetrised" equations. Geometrically, the reason why only half the equations on S_4 are required is that we need to choose half the points (less one!) from the lattice square $0 \leqslant \lambda_r \leqslant 2m+1$; the line S_4 is a diagonal of the square, and is "shared" between two triangles.

I showed that the covariance conditions (iv) and (v) reduced to a geometrical condition on the lattice, called the "rectangle rule"; it states that if $\underset{\sim}{\lambda}$ is a point of Q_m then all points in the rectangle R of Figure 2, with diagonal running from $\underset{\sim}{0}$ to $\underset{\sim}{\lambda}$, must be in the set Q_m. This does not of itself imply that the boundary of Q_m (excluding the axes) should be a line making an angle $\frac{1}{4}\pi$ with the axes. However, if $\underset{\sim}{\lambda}$ is a "symmetrised point", contributing to a symmetrised equation, no other points in the rectangle R may be symmetrised points. So all symmetrised points must lie on a single line making an angle $\frac{1}{4}\pi$ with the axes, and if symmetrised equations are adopted, one is forced to choose the triangular configuration of Figure 1. The fact that Q_m may not contain points (2m+1,0) and (0,2m+1) means that the triangle can be no larger, and that the configuration is unique.

When I showed this solution to Peter Graves-Morris and Dick Hughes Jones, they quickly pointed out that the constants A_1, A_2 in any transformation (4) had to be equal if the form of the symmetrised equations (from S_4) were to be preserved, and that a change of scale of only one variable changed the relative weights of the points (7) in these equations. My choice of equal weights was therefore arbitrary; this problem of weighting has been fully studied, and I shall discuss it later.

Two additional properties of the approximants I proposed were established[5] by Alan Common and Peter Graves-Morris:

(viii) Unitarity Property

The (diagonal) approximants of a unitary function are unitary. This property is derived directly from reciprocal covariance, as for Padé approximants, and is of particular importance in quantum scattering theory.

(ix) Factorisation Property

The [m/m] approximant to a function of the form $f(z_1)g(z_2)$ is $f_{m/m}(z_1)g_{m/m}(z_2)$.

John Gammel pointed out a further property[6]:

(x) Addition Property

The [m/m] approximant to a function of the form $f(z_1)+g(z_2)$ is $f_{m/m}(z_1)+g_{m/m}(z_2)$.

The last two properties, like reciprocal and homographic covariance and the projection property, are important in showing that the approximants behave in many respects like the functions they approximate.

In the paper that I published[2], I wrote out a list of possible generalisations and further problems, and a group of staff began to work on several of these problems. Alan Genz began to study methods of acceleration of convergence of double sequences, some of them suggested by the two-variable approximation, and he is reporting some of his results at this colloquium. Dick Hughes Jones, Peter Graves-Morris and Gordon Makinson (working on his own initially) began to study the algebraic properties of the linear equations, while John McEwan and I looked for a generalisation of the diagonal approximants to three variables and N variables, imposing the same geometrical conditions, notably the "rectangle rule".

John McEwan became involved in the work when I explained to him the geometrical nature of the problem of diagonal approximants for triple series. In our search for the 3-dimensional configuration, he and I first thought of a rather complicated configuration. I asked my daughter Carol to make a model of the volume proposed; next morning, when she had done this, it was evident that the rectangle rule was violated, and we then thought of a much simpler volume region, which seemed to be right. Again I asked Carol to make a model of this solid, explaining that its volume was twice the volume m^3 of the cube shown in Figure 3. Next morning she had made four of these models, and pointed out that they fitted together exactly to make a cube of volume $(2m)^3$; two of these solids are shown in Figure 3. We still had to solve the symmetrisation problem; after much thought and discussion, John McEwan came up with the solution of double symmetrisation over the three surfaces S_4 shown in Figure 4, and of triple symmetrisation over points on the three edges where these surfaces met. The reason for this symmetrisation becomes clearer when one sees the four volume elements fitted together to form the larger cube; each triangular surface is shared between two volume elements, while each edge is shared between three volume elements. We had to write down the sets of linear equations in concise algebraic notation, and found that it was remarkably simple to do; further, I realised that the partition of numbers of equations between volumes, surfaces, and edges followed a very simple rule which could easily be generalised to systems with an arbitrary number N of variables. In this way it turned out to be relatively easy to generalise the scheme to power series in N variables. The geometrical structures of Figures 1 and 4 thus generalise to N dimensions.

In Figure 4, if one looks at the points with one component of $\underset{\sim}{\lambda}$ zero (on the rear vertical plane, say), we see that they form exactly the configuration of Figure 1, with the correct double symmetrisation. These points correspond to the equations obtained when one variable is put zero, so that the three-variable (and likewise the N-variable) approximants satisfy the projection property. The system of approximants thus forms an infinite-dimensional space with a natural projection property. The fact that the whole space of approximants mimics the behaviour of the original many-variable series so well encourages our belief that the approximants will often represent functions of several variables well. This work on N-variable diagonal approximants was published[7] in 1974. During the same period, Gordon Makinson, Dick Hughes Jones and Peter Graves-Morris were studying the algebraic structure of the two-variable equations, defining off-diagonal generalisations, and writing programmes to carry out computations. They produced, over a short period of time, a series of three papers containing their results. The first paper[8], by Hughes Jones and Makinson, elucidated the algebraic structure of my two-variable equations; this structure turned out to be remarkably simple and convenient. Apart from the complication due to the symmetrised equations, the matrix which had to be inverted was of the lower block diagonal form

$$\begin{pmatrix} D_m^{(1)} & & & & & & \\ R_m^{(1)} & D_m^{(1)} & & & & & \\ \cdots & \cdots & \cdots & & & & \\ R_1^{(1)} & & D_1^{(1)} & & & & \\ & & & D_m^{(2)} & & & \\ & & & R_m^{(2)} & D_m^{(2)} & & \\ & & & \cdots & \cdots & \cdots & \\ & & & R_1^{(2)} & & D_1^{(2)} & \end{pmatrix}, \tag{8}$$

where $D_1^{(1)},\ldots,D_m^{(1)}$ and $D_1^{(2)},\ldots,D_m^{(2)}$ are just the square matrices which have to be inverted in order to calculate the first m Padé approximants of the single power series $f(z_1,0)$ and $f(0,z_2)$. Thus the inversion of a matrix of dimension (m^2+2m) is essentially reduced to the calculation of two sequences of diagonal Padé approximants up to order m. Further, it was now clear that the matrix in general had a unique inverse, and that existence and uniqueness theorems could be established. The structure of the equations was interpreted in terms of "prongs"; the L-shaped set of points in Figure 1 is a "prong", and is defined by the point (p,p) on the line of symmetry of the lattice region; it defines a set of equations which are considered as a block. The fact that the number of equations corresponding to points of a prong inside the square is equal to the number corresponding to points outside is directly related to the fact that the block matrices on the diagonal of the matrix (8) are square matrices, ensuring that the matrix (8) can generally be inverted.

In the second paper[9], Graves-Morris, Hughes Jones and Makinson used the prong method to define symmetric off-diagonal approximants (S.O.D's); this method ensures that the linear equations are generally soluble, and uniquely defines approximants symmetric in two variables which have properties analogous to those of off-diagonal Padé approximants, including reciprocal covariance; they also satisfy the projection property. The first computations using diagonal approximants and S.O.D's, calculating the beta-function and approximating its singularities, were reported in this paper; the work provided welcome confirmation that the approximants did in fact approximate! An ALGOL procedure[10] was published along with this paper. Figure 5 shows the lattice configuration corresponding to the S.O.D. linear equations, and typical prongs.

The third paper[11], by Dick Hughes Jones, extended the prong method to series with N variables; for diagonal approximants, the natural extension of 2-variable approximants using this approach turned out to be exactly the approximants defined by John McEwan and I; this meant that these approximants satisfied all the properties (i)-(vi) and (viii)-(x), and also had good algebraic properties. The paper also extended S.O.D's to N-variable series, and defined general off-diagonal approximants (G.O.D's), in which the maximum powers of all N variables in both numerator and denominator were arbitrary; the G.O.D's were therefore the most general approximants possible. The G.O.D's again satisfy the fundamental properties similar to those of the diagonal approximants, excluding of course homographic covariance. The lattice point configurations for three-dimensional S.O.D's and for two- and three-dimensional G.O.D's are shown in Figures 6, 7 and 8. In Figure 6, the two types of prong needed to define the system of equations are exhibited, and a typical 2-dimensional prong for a G.O.D. is shown in Figure 7.

In the early summer of 1973, Peter Graves-Morris and I both began thinking about convergence theorems, and decided to collaborate in trying to generalise the theorem of de Montessus de Ballore. This involved simplifying and generalising Gragg's proof of de Montessus' theorem for simple poles, in order to provide a simple enough starting point. The work took nine months to complete, and we had to do a great deal of hard classical analysis and algebra of determinants; we each contributed several crucial ideas, and I do not think that either of us would have completed the work alone. We succeeded in establishing some limited generalisations of de Montessus' theorem to two-variable approximants[12]; the variety and complexity of functions of two variables makes it necessary to impose conditions which are not needed for a single variable. There appears to be no reason why our theorems cannot be generalised straightforwardly to sequences of N-variable approximants.

In this paper, we introduced arbitrary weights into the symmetrised equations, since this arbitrariness affected none of the properties of the whole scheme. The problem of weighting had been with us from the beginning, and in the autumn of 1974, while he was working at Brookhaven N.L., Peter Graves-Morris wrote telling us of a proposal to choose the weights in order to maximise the denominator determinant in solving the linear equations[13]; it occurred to me that this choice might also ensure convariance of the equations under relative scale transformations. However, Dick Hughes Jones and I quickly discovered that covariance was ensured by exactly the *inverse* choice of weights[14]; we therefore had solved the problem of providing full covariance of diagonal approximants under the group (4) with r=1,2,...,N, for any N. This choice, nevertheless, courts numerical disaster because it may help to make the equations almost degenerate. Peter Graves-Morris and David Roberts have compared numerically[15] the various choices of weighting factors; there is not a great deal of difference, but PETCH (Peter's choice) appears to give rather more consistent results than SCINCH (scale invariant choice). In their paper[13], Graves-Morris and Hughes Jones have analysed possible degeneracies of G.O.D. two-variable approximants, showing that new types of degeneracies arise, compared with Padé approximants.

Two applications to problems in physics have been made. Peter Graves-Morris and his student Charles Samwell have applied two-variable approximants to the study of problems in potential theory[16]; this work is being reported at Marseille later in the week. My own view is that these examples, plus one by John Gammel, are very encouraging. The other application arose out of a lecture I gave to the Nottingham University Student Mathematical Society; I was talking on Padé approximants, but could not resist mentioning N-variable approximants at the end. Dr. David Wood (of Nottingham) was present, and immediately proposed using these approximants on several double series arising in critical phenomena. The first results from Nottingham[17,18] indicate that the new approximants solve some problems not previously solved. Through David Roberts, we supplied programmes[10,19] for this work, and he has collaborated in the study of a number of numerical examples[20], which again give encouraging results. The scope for applying this technique to double (and, later, triple) power series is enormous, and at this conference we have heard of problems in fluid mechanics which give rise naturally to double series. Theoretical chemistry is another field in which series in several variables arise naturally. While no technique can be expected to solve all problems to which it can in theory be applied, it seems that we are just at the beginning of the investigation of the applicability of this N-variable approximation method to a wide variety of problems arising in many different fields.

There have been two other investigations of approximants of this general type, by Levin[21], and by Lutterodt and John[22,23]. These studies were framed more generally than those carried out in Canterbury, and were not directed at defining approximants satisfying the very specific properties which I originally sought to satisfy. Two other approaches to defining many-variable approximants have been studied : Alabiso and Butera have defined approximants through the two-variable moment problem[24], and in an excellent doctoral thesis[25], O'Donohoe has defined N-variable generalisations of continued fractions, not only matching a power series but also solving the interpolation problem. Both of these approaches give approximants differing from those I have described; it may be that three topics which overlap considerably for one-variable series, namely Padé approximation, continued fractions and the moment problem, have generalisations for many variables which are essentially different.

One factor which has been of great importance throughout this work has been the readiness of each member of our group in Canterbury to communicate and share his ideas; this was undoubtedly a major factor in the rapid development of the theory, and I would like to acknowledge the unselfish co-operation of all those connected with the work. I would also like to thank Sandra Bateman for her consistently excellent work in producing this and many other more complicated papers, often under serious pressure of time.
I am grateful to Dick Hughes Jones for allowing me to copy or adapt a number of his very clear diagrams, and to my daughter Carol for drawing some of the diagrams.

REFERENCES

1. J.S.R. Chisholm, Proceedings of Third International Colloquium on Advanced Computing Methods in Theoretical Physics (Marseille), ed. A. Visconti (1974), C-XIV-1.

2. J.S.R. Chisholm, Math. Comp. 27 (1973), 841.

3. J.L. Gammel, "Padé Approximants and Their Applications", ed. P.R. Graves-Morris (Academic Press, 1973), 3.

   C. Critchfield, Rocky Mountain J. Math. 4, 2 (1974).

4. J. Nuttall, J. Math. Anal. Applic. 31 (1970), 147.

5. A. K. Common & P. R. Graves-Morris, J. Inst. Math. Applic. 13 (1974), 229.

6. J.L. Gammel (Private Communication).

7. J.S.R. Chisholm & J. McEwan, Proc.Roy.Soc. A336 (1974), 421.

8. R. Hughes Jones & G.J. Makinson, J.Inst.Math.Applic. 13 (1974), 299.

9. P.R. Graves-Morris, R. Hughes Jones & G.J. Makinson, J.Inst. Math.Applic. 13 (1974), 311.

10. P.R. Graves-Morris, R. Hughes Jones & G.J. Makinson, Comp.J 18 (1975), 81.

11. R. Hughes Jones, Kent preprint (1973), submitted to J.Approx. Theory.

12. J.S.R. Chisholm & P.R. Graves-Morris, Proc.Roy.Soc. A336 (1974), 421.

13. P.R. Graves-Morris & R. Hughes Jones, Brookhaven N.L. preprint AMD 674 (1974).

14. J.S.R. Chisholm & R. Hughes Jones, Proc. Roy.Soc. (1975), A344, 365.

15. P.R. Graves-Morris & D.E. Roberts, Kent preprint (1975).

16. P.R. Graves-Morris & C.J. Samwell, Kent preprint (1975).

17. D.W. Wood & H.P. Griffiths, J.Phys.A. (Letters) 7(1974), L101.

18. P. Fox & D.W. Wood, J.Phys.A (1975), to be published.

19. P.R. Graves-Morris & D.E. Roberts, Comp.Phys.Comm., 9(1975), 46-50.

20. D.E. Roberts, D.W. Wood & H.P. Griffiths, J.Phys.A (1975), to be published.

21. D. Levin, Tel-Aviv preprint (1973).

22. C. Lutterodt, J.Phys.A, 7 (1974), 1027.

23. G. John & C. Lutterodt, submitted to Nuovo Cimento (1974).

24. C. Alabiso & P. Butera, J.Math.Phys. 16, 4 (1975), 840.

25. M.R. O'Donohoe, Ph.D. thesis (Brunel University, 1974).

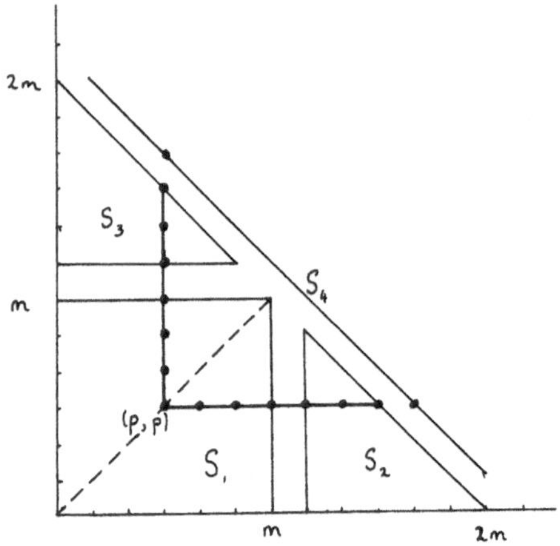

Fig.1 Parameter regions : 2-variable diagonal

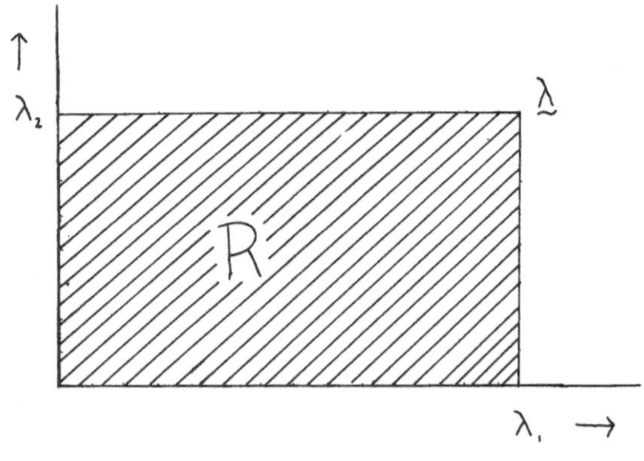

Fig.2 Rectangle rule : 2-variable

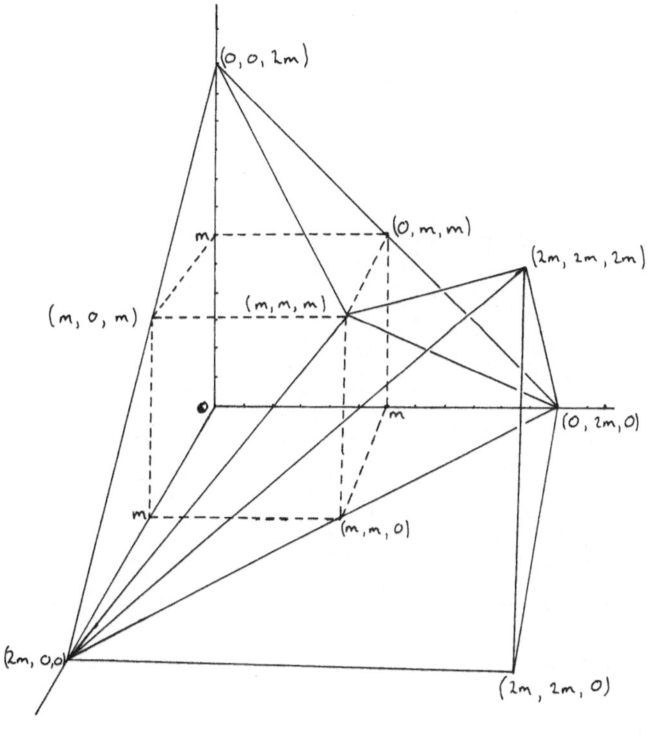

Fig.3 Volume regions : 3-variable

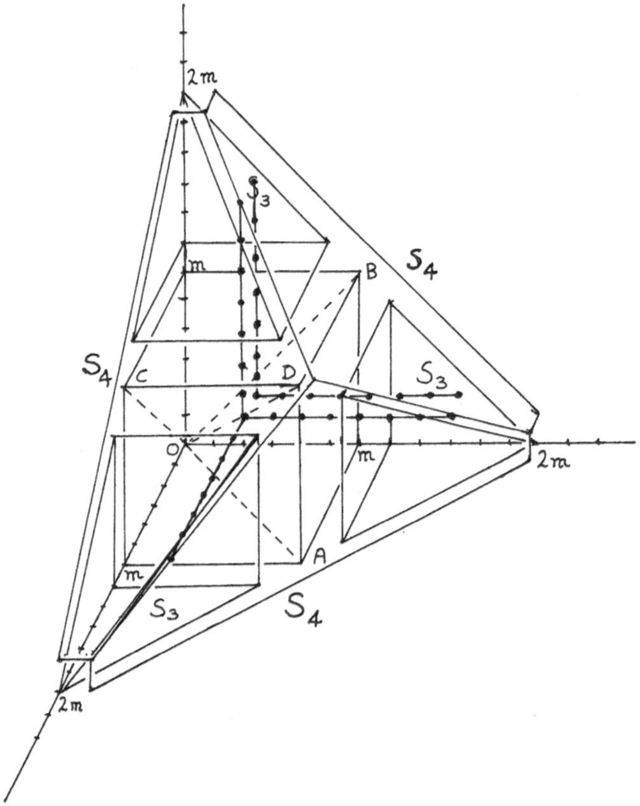

Fig.4 Parameter regions : 3-variable diagonal

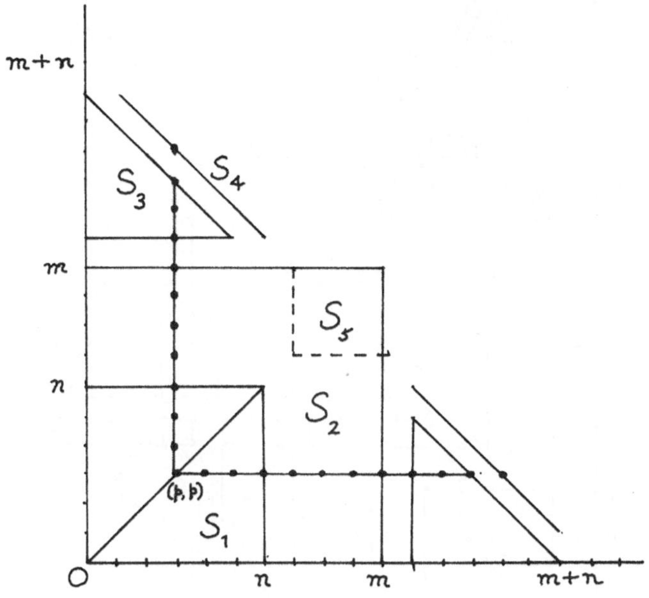

Fig.5 Parameter regions : 2-variable S.O.D.

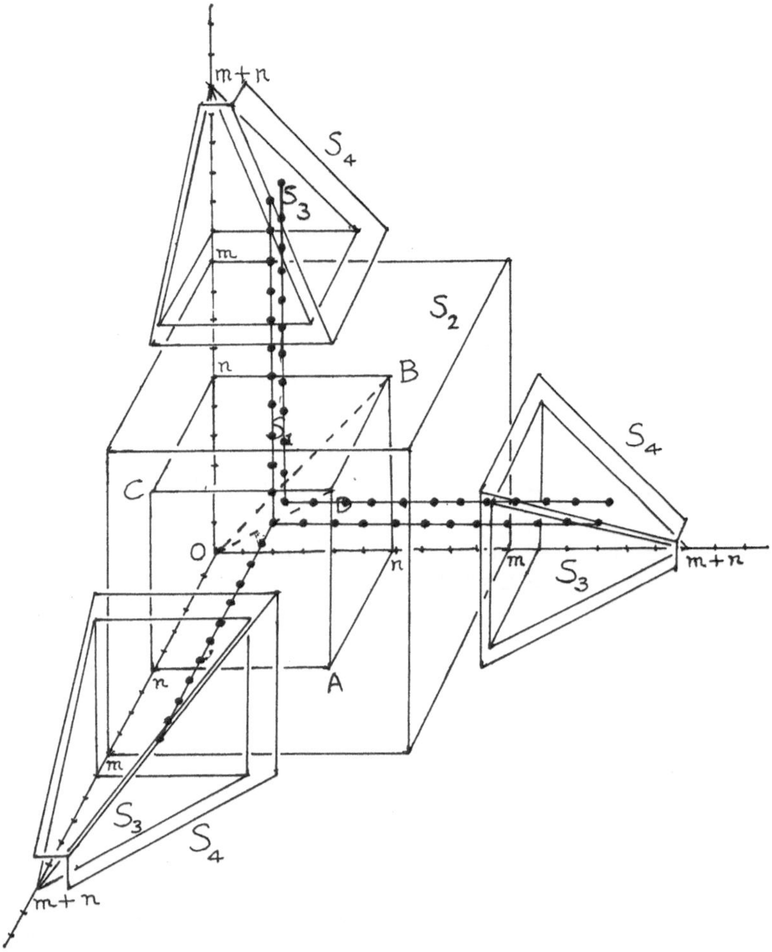

Fig.6 Parameter regions : 3-variable S.O.D.

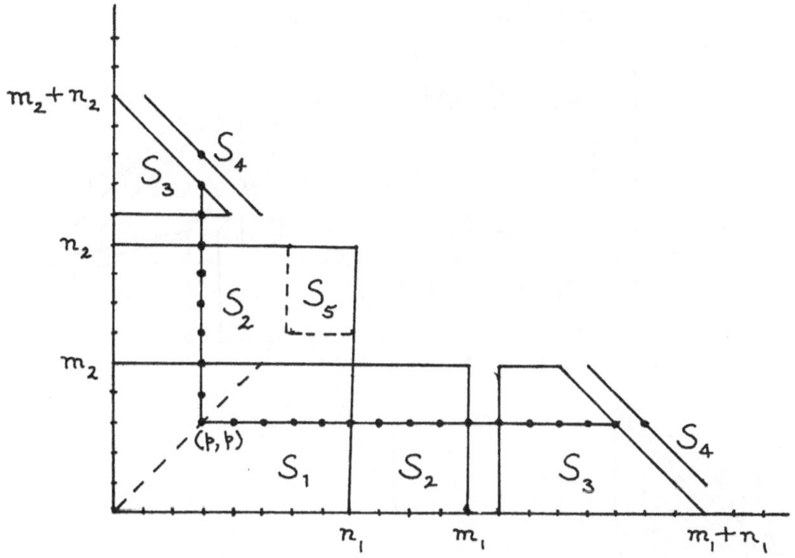

Fig.7 Parameter Regions : 2-variable G.O.D.

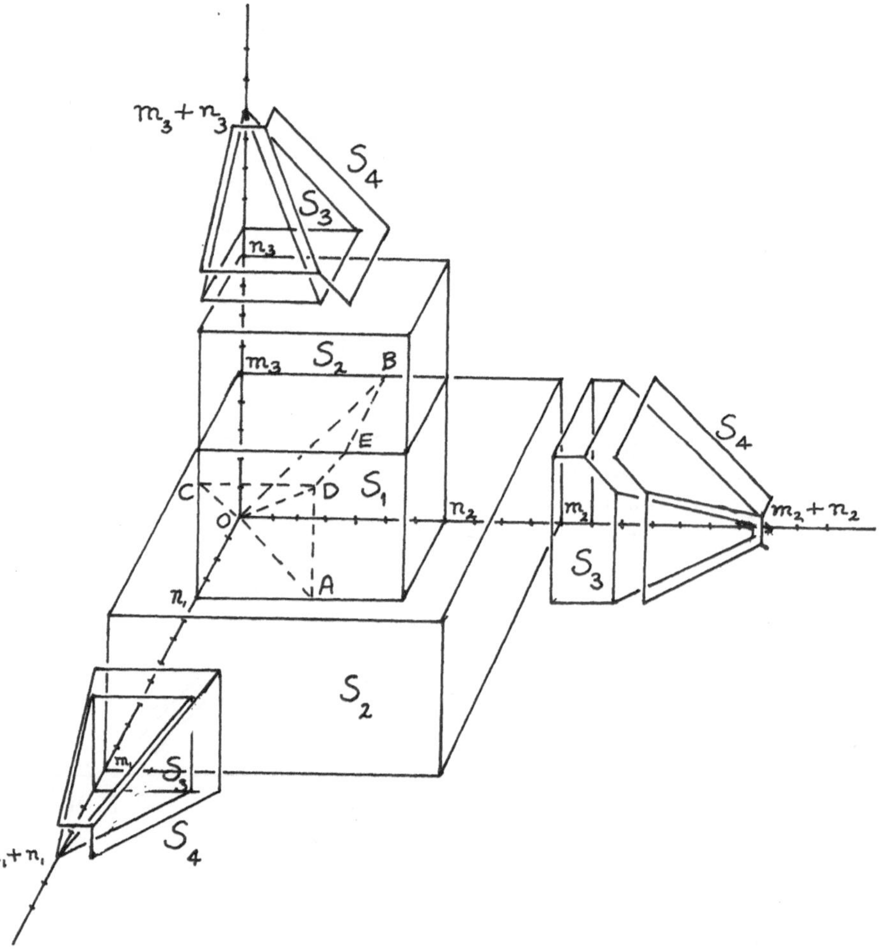

Fig.8 Parameter regions : 3-variable G.O.D.

CONVERGENCE OF ROWS OF THE PADE TABLE

P. R. Graves-Morris
Mathematical Institute
University of Kent
Canterbury
Kent
England

Abstract

Some of the convergence theorems about rows of the Padé table of analytic functions are reviewed, especially Beardon's theorem and de Montessus' theorem. The progress on convergence theorems for the third row, the "poles out" theorem, and de Montessus' theorem for two variables are explained. Two conjectures about convergence of rows, one of which is a counterpart of the conjecture of Baker, Gammel and Wills for diagonal sequences, are boldly made.

Introduction

The ideal introduction is Chapter 11 of "The Essentials of Padé Approximants" [1]. However, the pertinent facts may be selected and the tale is not too complicated. It starts by defining Padé Approximants, P.A's for short, by $f_{L/M}(z) = A(z)/B(z)$ where $A(z)$ is a polynomial of degree at most L, $B(z)$ is a polynomial of degree at most M, $B(0)=1$ and the Maclaurin expansion of $f_{L/M}(z)$ agrees with that of $f(z)$ up to and including the coefficients of z^{M+N}.

But before commencing on the theorems and their justifications, it is as well to reconsider some of the motives for using the Padé table. Given a formal power series, how does one reconstruct the function it represents? There is little else one can do except form the Padé table and inspect some suitable sequences. If the first row converges, which means that the sequence of [L/0] approximants converges, which is the same as saying that the Taylor series converges, then one is either in or on the circle of convergence of the given power series and there is no difficulty in principle. In practice, convergence may be too slow for the method to be of value. One of the methods of accelerating convergence is Aitken's δ^2 method, which turns out to be the same as using the second row of the Padé table, which is the sequence of [L/1] approximants. If this converges, again the problem is solved in principle. Likewise, convergence can be accelerated again by looking at the third row, fourth row et cetera. What have we gained? In practice, the answer is that convergence has been greatly accelerated, and going not too far down the third row gives a good answer if convergence of the first row is slow. What has been lost? You never get something for nothing, and the price of using any row except the first is that holomorphic functions are known to exist for which an infinite subsequence of approximants does not converge. The price is the loss of the guarantee of convergence of the Taylor series of holomorphic functions.

The natural suggestion prompted by the foregoing is the use of the diagonal sequence. Padé folklore is that you should normally use the diagonal sequence, and I entirely support this popular movement. Various reasons support the use of the sequence of diagonal approximants in cases where there is no information to suggest an alternative. I propose that the following reason is as strong as any other. One conjectures that the given power series is the expansion of a meromorphic function f(z). Thus $\{f(z)\}^{-1}$ is also a meromorphic function and the diagonal sequence of P.A's to f(z) is the only simple sequence which treats f(z) and $\{f(z)\}^{-1}$ symmetrically. These remarks are based on the theorem that if $g(z) = \{f(z)\}^{-1}$, then $g_{L/L}(z) = \{f_{L/L}(z)\}^{-1}$. There is also considerable numerical experience (by authors who use sufficient numerical accuracy to give credible results) to back the choice of diagonal approximants. (Other authors quote the theorem of Baker, Gammel and Wills [2] as evidence to support the use of diagonal P.A's, but I think the argument is misleading because so many functions have essential singularities at infinity). Of course, the fly in the ointment is Gammel's example of a holomorphic function for which an infinite subsequence of diagonal approximants diverge [3]. The reason for mentioning the importance of diagonal approximants is to emphasise the importance of proving the conjecture of Baker, Gammel and Wills, and equally to state the importance of discovering the class of functions (which certainly includes the class of Stieltjes functions) for which the diagonal sequence of P.A's converges. The answers are not going to be easy to find, and, as a start, it is easier, but not easy, to consider row sequences. So one motive for studying row sequences is to learn something about how to treat diagonal sequences.

The first and second rows

There is little to say about the first row except that it is the set of truncated Maclaurin series.

The second row is much more interesting because it appears to be typical of what is expected, but not yet established, for the general row. Everything is contained in Beardon's theorems [4] which have the following implications:-

Let f(z) have radius of convergence R. Let the poles of the [L/1] approximants be at $z=p_L$. Then

$$\underline{\lim}\, p_L \leq R \leq \overline{\lim}\, p_L \quad .$$

It follows, as this suggests, that an infinite subsequence of [L/1] approximants exists which converge uniformly to f(z) on $|z|\leq\rho$ for any $\rho<R$. It follows, as the inequality suggests, that functions analytic in $|z|<R$ may be found for which an infinite subsequence of [L/1] approximants do not converge at any point in $|z|<R$, or even on a point set which is dense in $|z|<R$.

The proofs are quite simple and follow from the fact that p_L is given by the ratio of successive terms of the Maclaurin series of f(z). The final remark about divergence on a dense point set follows from Perron's example [5].

de Montessus' theorem

Let f(z) be meromorphic in $|z|<R$ with precisely M poles in $|z|<R$, counting multiplicity. Then $f_{L/M}(z) \to f(z)$ as $L\to\infty$ uniformly on $|z|<\rho$ for any $\rho<R$, except on arbitrarily small open neighbourhoods of the poles.

This means that f(z) has M poles in $|z|<R$, where double poles count double, and p^{th} order poles count p times. The important thing is that, given R, we must know M. Once the radius of meromorphy, R, and the number of poles enclosed, M, are known, de Montessus' theorem establishes as much convergence as one can possibly expect,

and is a very powerful theorem. The old proof [6] is now superseded by Saff's elegant proof [7]. From a constructive viewpoint, the work of Gragg [8] and simplified by Chisholm and Graves-Morris [9] has some advantages. These last three authors show that the M dimensional Hankel determinants of the coefficients approach limits determined by the M poles of the given function nearest the origin and their residues. The advantage of Saff's proof is that the case of multiple poles need not be treated as a special case.

The drawbacks of de Montessus' theorem and the need for other results are most simply seen by a few examples. Suppose f(z) is holomorphic. Then de Montessus' theorem only applies to the first row, which is the sequence of [L/0] approximants. Or alternatively, suppose that we know that f(z) has at least M poles. Does it follow that [L/M] approximants converge in some domain as $L\to\infty$? The answer is no, not necessarily. Suppose the poles of f(z) are ordered according to their distance from the origin, and the $M-1^{th}$, M^{th} and $M+1^{th}$ poles are equidistant from the origin. Then there is no circle, centre $z=0$, which contains precisely M poles of the function and de Montessus theorem does not apply in this case.

Existence of Approximants

It is worth mentioning that the apparently harmless definition of P.A's stated in the introduction is, in fact, the modern or Baker definition [10]. It has several important consequences. The The first is that if approximants cannot be found to satisfy the conditions, then they are declared not to exist. The best known example is the non-existence of a [1/1] approximant to $1+z^2$. The idea that interpolatory rational fractions do not exist in certain circumstances is well known, and one advantage of the Baker definition is that it does not obscure the problem by introducing deficiency indices or by cross-multiplying by zero or by any other subterfuge. The statement that an infinite subsequence of approximants of any row of the Padé table exists is then a non-trivial theorem [3], and is true for any formal power series. Nothing follows about convergence of the extant approximants to any limit function.

The Third Row

The principal new result I wish to mention is a convergence theorem for the third row of the Padé table. Baker and I [11] have established that at least an infinite subsequence of [L/2] approximants of a holomorphic function converge to the function uniformly in any compact region of the complex plane.

Proof The proof is complicated, but it is quite easy to outline the principal ideas. Suppose that the given holomorphic function is

$$f(z) = \sum_{i=0}^{\infty} c_i z^i .$$

Set up an array of vectors $\underline{v}_i = (c_i, c_{i+1})$ in the space $\mathbb{C}^2$, and $|\underline{v}_i| = v_i = \{|c_i|^2 + |c_{i+1}|^2\}^{\frac{1}{2}}$. It is quite easy to show that, for any positive X no matter how large, there exists an infinite subsequence of ratios $R_i = v_i/v_{i+1}$ which are greater than X. For otherwise we can show that f(z) has radius of convergence equal to X, which is untrue by hypothesis.

Second, we may examine the denominators of the [L/2] Padé approximants, which are

$$Q_L(z) = 1 + \alpha_L z + \beta_L z^2 .$$

Provided that an infinite subsequence of values of L may be found so that $|\alpha_L|$ and $|\beta_L|$ are sufficiently small, then $Q_L(z)$ has no zeroes in the compact region where the approximants are expected to converge.

Since

$$\alpha_L = \begin{vmatrix} c_{L-1} & c_{L+1} \\ c_L & c_{L+2} \end{vmatrix} \Big/ \begin{vmatrix} c_{L-1} & c_L \\ c_L & c_{L+1} \end{vmatrix}$$

and

$$\beta_L = \begin{vmatrix} c_L & c_{L+1} \\ c_{L+1} & c_{L+2} \end{vmatrix} \Big/ \begin{vmatrix} c_{L-1} & c_L \\ c_L & c_{L+1} \end{vmatrix}$$

we see the condition for α_L and β_L to be small is that $\begin{vmatrix} c_{L-1} & c_L \\ c_L & c_{L+1} \end{vmatrix}$ is not anomalously small. This would only be true if $\underline{v}_{L-1}$ and $\underline{v}_L$ were almost parallel, and in that case, the [L-1/2] approximant turns out to be a hopeful candidate. The proof now becomes more complicated, but runs along the lines that if R_L is sufficiently large, the [L/2] approximant is superficially a good approximation to f(z). The only problem occurs when the denominator determinant is small, in which case the [L-1/2] approximant is superficially a good approximant. We proceed by induction. Either a good approximant is found, or else a whole sequence of determinants are small, which means that the function closely approximates a geometric function, which is not holomorphic. Thus there must exist an infinite subsequence of convergent [L/2] approximants.

This is only the outline of the proof. The actual proof removes the condition in Baker's earlier result [3] that f(z) had to be a holomorphic function of order less than one. However, our theorem does not go as far as we would like. The best theorem is expected to require only that f(z) have a circle of convergence, and to prove that an infinite subsequence of [L/2] approximants converge to f(z) within that circle. That remains to be established.

A Conjecture about Convergence of Rows

The foregoing remarks led Baker and me to make the following conjecture [11]:

"At least an infinite subsequence of [L/M] approximants of a function converges to the function within the largest circle, centred on the origin which contains not more than M poles of the given function and within which the function is meromorphic".

We can establish this result in various special cases which give credibility to the conjecture. For a start, if the circle contains precisely M poles, the theorem follows from de Montessus' theorem. If the given function is holomorphic, the result of the previous theorem proves the conjecture for the third row. If the function is analytic within the circle, then Beardon's theorem establishes our conjecture for the second row. Our conjecture is trivially true for the first row. There remains the question about fourth and lower rows and also the question of functions meromorphic in the circle and with precisely one pole in the circle. The latter possibility we deal with next.

A "pole out" theorem

Suppose the function c(z) is given, which is meromorphic in a circle Γ_R of radius R, centred on the origin, and has m poles, counting multiplicity within Γ_R. Let $\sigma(z)$ be the monic polynomial of degree m for which both $g(z) = \sigma(z)\ c(z)$ and $h(z) = \sigma(z)\ g(z)$ are analytic in Γ_R. Suppose that a sequence $S = \{L_1, L_2, \dots\}$ is given such that $[L_i/1]_h$ converges to h(z) in Γ_R as $i \to \infty$. Then $[L_i'-m/m+1]$ converges to c(z) for some infinite subsequence $S' = \{L_1', L_2', \dots\} \subset S$ on any compact set $\mathfrak{D}$ satisfying

$$\mathfrak{D} \subset \{z\colon \sigma(z) \neq 0 \text{ and } |z| < R\} \ .$$

What this theorem means is that we have established a link between convergence of [L-m/m+1] approximants to the meromorphic function c(z) and the [L/1] approximants to the analytic function $h(z) = \{\sigma(z)\}^2 c(z)$. It is a little surprising that the link is with h(z) rather than $g(z) = \sigma(z)\, c(z)$. At any rate, with m=1, this theorem establishes our previous conjecture about rows as far as the third row for functions meromorphic in a region with at least one pole.

For the proof of the previous theorem, we refer to our preprint [11]. In outline, we note that, for m=1,

$$Q^{[L-1/2]}(z) = \begin{vmatrix} z^2 & z & 1 \\ c_{L-2} & c_{L-1} & c_L \\ c_{L-1} & c_L & c_{L+1} \end{vmatrix} .$$

Let
$$\sigma(z) = z + \sigma_1$$
$$g(z) = \sum_{j=0}^{\infty} g_j z^j$$
$$h(z) = \sum_{j=0}^{\infty} h_j z^j .$$

Then
$$g_j = c_{j-1} + \sigma_1 c_j$$
$$h_j = g_{j-1} + \sigma_1 g_j$$

and

$$Q^{[L-1/2]}(z) = \begin{vmatrix} z\sigma(z) & \sigma(z) & 1 \\ h_L & h_{L+1} & g_{L+1} \\ g_L & g_{L+1} & c_{L+1} \end{vmatrix}$$

$$= c_{L+1} \det \left\{ \begin{pmatrix} z\sigma(z) & \sigma(z) \\ h_L & h_{L+1} \end{pmatrix} - \begin{pmatrix} e_{11} & e_{12} \\ e_{21} & e_{22} \end{pmatrix} \right\}$$

where

$$\begin{pmatrix} e_{11} & e_{12} \\ e_{21} & e_{22} \end{pmatrix} = \begin{pmatrix} 1 \\ g_{L+1} \end{pmatrix} c_{L+1}^{-1} (g_L \;\; g_{L+1}) .$$

Then it turns out that $Q^{[L-1/2]}(z)$ is dominated by the term $-c_{L+1}\, h_L\, \sigma(z)$, if L is chosen so that $|h_L/h_{L+1}|$ is suitably large. This is not surprising, since c_{L+1} is dominated by the contribution of the pole of c(z) and h_L and g_L are expansion coefficients of holomorphic functions.

Once again, the foregoing remarks lead to a conjecture, which we entitle the role of the poles conjecture.

"Suppose that a function c(z) is given, which is meromorphic in a circle Γ_R of radius R, centre the origin, and has m poles, counting multiplicity, within Γ_R. Let $\sigma(z)$ be the monic polynomial of degree m for which $g(z) = \sigma(z)\, c(z)$ and $h(z) = \sigma(z)\, g(z)$ are analytic in Γ_R. Suppose that an infinite sequence $S = \{L_1, L_2, \ldots\}$ is given such that $[L_i/\mu]_h$ converges to h(z) in Γ_R. Then we conjecture that $[L_i'-1/\mu+1]_c$ converges to c(z) for some infinite subsequence $S' = \{L_1', L_2', \ldots\} \subset S$ on any compact set $\mathfrak{D}$ satisfying

$$\mathfrak{D} \subset \{z: \sigma(z) \neq 0,\ |z| < R\}\ .$$

Baker and I prove in our preprint that this result holds also in the case of $\mu=2$ and h(z) being a holomorphic function of order less than 1.

<u>de Montessus' Theorem in Two Variables</u>

The final topic I wish to mention is the analogue, due to Chisholm and myself [9], of de Montessus' theorem in convergence of Canterbury Approximants. We define a C.A. in this context by

$$f_{m_1 m_2/n_1 n_2}(x,y) = \frac{\sum_{i=0}^{m_1} \sum_{j=0}^{m_2} a_{ij} x^i y^j}{\sum_{i=0}^{n_1} \sum_{j=0}^{n_2} b_{ij} x^i y^j} = \frac{a(x,y)}{b(x,y)}\ .$$

In this expression, a(x,y) is a polynomial of maximum degree n_1 in x and maximum degree n_2 in y and the set $\{a_{ij}\}$ occupies a rectangular block of dimensions $(m_1+1)\times(m_2+1)$ in the (i,j) lattice space. Similarly, the denominator is a polynomial and $\{b_{ij}\}$ occupies a $(n_1+1)\times(n_2+1)$ block of lattice space. We define $b_{oo}=1$, so that there are $(m_1+1)(m_2+1) + (n_1+1)(n_2+1) - 1$ coefficients to be determined. Suppose that we seek approximants to

$$f(x,y) = \sum_{i=o}^{\infty} \sum_{j=o}^{\infty} c_{ij}x^i y^j \quad .$$

Then we cross multiply to give

$$\left(\sum_{i=o}^{n_1} \sum_{j=o}^{n_2} b_{ij}x^i y^j\right)\left(\sum_{i=o}^{\infty} \sum_{j=o}^{\infty} c_{ij}x^i y^j\right) - \sum_{i=o}^{m_1} \sum_{j=o}^{m_2} a_{ij}x^i y^j$$

$$= \sum_{i=o}^{\infty} \sum_{j=o}^{\infty} d_{ij}x^i y^j \ .$$

We require as many of the d_{ij} to be zero as possible, and look at the d_{ij} lattice space to see what the equations should be. First we need the rectangular block $(0\leqslant i\leqslant m_1) \otimes (0\leqslant j\leqslant m_2)$ which determines the $\{a_{ij}\}$, once the b_{ij} are known. Second, we need $(n_1+1)\times(n_2+1) - 1$ further equations. Without going into details, this second block is divided into a triangular region and a trapezoidal region, and these are appended to the first block to give the correct number of self-consistent equations for $\{a_{ij}\}$ and $\{b_{ij}\}$. The details are explained in [9,12,13].

The approximants so defined to have a variety of useful properties, such as reduction to Padé Approximants and factorisation, as is explained in Prof. Chisholm's contribution to this book. In seeking to generalise a theorem about the convergence of rows of the Padé table to meromorphic functions, we seek a generalisation of the notion of rows and of meromorphic functions. A meromorphic function of several variables is a section of a sheaf of germs [14]. Such precision is necessary because complex functions of two or more variables are much more complicated than in the one variable case, and great caution is required. With an eye on the possible, we state a theorem about functions of the form $f(x,y) = A(x,y)/B(x,y)$, where $A(x,y)$ is analytic in a domain $|x| \leq R_1$, $|y| \leq R_2$ to be specified and $B(x,y)$ is a polynomial:

$$B(x,y) = \sum_{i=0}^{n_1} \sum_{j=0}^{n_2} B_{ij} x^i y^j \quad \text{with} \quad B_{oo} = 1.$$

To ensure that it is genuinely a polynomial of degree $n_1 \times n_2$, we require that $B(x,0)$ has n_1 zeros, counting multiplicity, at $p_\nu^{(1)}$ $(\nu=1,2,\ldots,n_1)$ which are ordered so that $0 < |p_1^{(1)}| \leq |p_2^{(1)}| \leq \ldots \leq |p_{n_1}^{(1)}| < R_1$ and similarly that $B(0,y)$ has n_2 zeros, counting multiplicity, at $p_\nu^{(2)}$ $(\nu=1,2,\ldots,n_2)$ which are ordered so that $0 < |p_1^{(2)}| \leq |p_2^{(2)}| \leq \ldots \leq |p_{n_2}^{(2)}| < R_2$. We further require, for reasons which are by no means immediately obvious, that $f(x,0)$ have no poles equidistant from the origin except multipoles, and a similar requirement for $f(0,y)$. In equations, this means that if $|p_{\nu-1}^{(i)}| = |p_\nu^{(i)}|$, then $p_{\nu-1}^{(i)} = p_\nu^{(i)}$ for $\nu=2,3,\ldots,n_i$ and $i=1,2$.

We have two reasonably strong theorems, which are:-

Theorem 1

Let $B(x,y) = g_1(x)\, g_2(y)$, so that $B(x,y)$ factorises. Let R_1, R_2 be any numbers for which $R_1 > |p_{n_1}^{(1)}|$ and $R_2 > |p_{n_2}^{(2)}|$. Then the C.A's $[m_1,m_2/n_1,n_2]$ converge uniformly to $f(x,y)$ as $\min(m_1,m_2) \to \infty$, on any compact subset of $\{x,y: |x| \leq R_1, |y| \leq R_2, B(x,y) \neq 0\}$.

Theorem 2

Let $f(x,y) = f(y,x)$, which means that $f(x,y)$ is symmetric. Then we may set $R=R_1=R_2$, $m=m_1=m_2$, $n=n_1=n_2$, $p_i=p_i^{(1)}=p_i^{(2)}$.

Let

$$R > |p_n| \prod_{i=1}^{n} |p_i/p_1|$$

which means that $A(x,y)$ has to be analytic in a larger domain than $(|x|<|p_n|) \otimes (|y|<|p_n|)$. Then the C.A's $[m,m/n,n]$ converge uniformly to $f(x,y)$ on any compact subset of $\{x,y: |x|\leq R, |y|\leq R, B(x,y)\neq 0\}$.

The proofs of these theorems are too long even to outline. There are also other theorems for similar cases. But the spirit is plain. We have analogues of de Montessus' theorem, but with several constraints on the nature and location of the singularities allowed.

Postscript

Theorems about rows of the Padé table have been discussed. I have preferred to term the set of [L/M] approximants with M fixed and L=0,1,2,... a row because the Taylor expansion is usually expressed along the line of writing; this reason is not too strong, and it may be that "Columns of the Padé Table" would have been a better title. Only time will tell which nomenclature is the more popular.

I suspect that the Baker Graves-Morris conjecture about convergence of subsequences of rows (or did I mean columns?) will be proved within a few years. Maybe the role of the poles conjecture (also called the "poles out" conjecture) will take longer. The discovery of convergence theorems for Canterbury Approximants is clearly incredibly difficult, being both a challenge of ingenuity and endurance.

REFERENCES

1. "The Essentials of Pade Approximants", G.A. Baker, Jr., Academic Press N.Y. (1975).

2. G.A. Baker, J.L. Gammel and J.G. Wills, J. Math. Anal. Appl. 31 147 (1970).

3. G.A. Baker, J. Math. Anal. Appl. 43 498 (1973).

4. A.F. Beardon, J. Math. Anal. Appl. 21 469 (1968).

5. O. Perron, "Die Lehre von den Kettenbrüchen", Vol.II p.270, B.G. Teubner (1957).

6. R. de Montessus de Ballore, Bull. Soc. Math. France 30 28 (1902).

7. E.B. Saff, J. Approx. Theory 6 63 (1972).

8. W.B. Gragg in "Padé Approximants and Their Applications", Academic Press (London) p.117 (1973).

9. J.S.R. Chisholm and P.R. Graves-Morris, Proc. Roy. Soc. A 342 341 (1975).

10. G.A. Baker in "The Padé Approximant in Theoretical Physics" eds. G.A. Baker, Jr. and J.L. Gammel, Academic Press, N.Y. (1970).

11. G.A. Baker and P.R. Graves-Morris, "Convergence of Rows of the Padé Table", Brookhaven preprint AMD 675 (1974).

12. R. Hughes Jones, "General Rational Approximants in N-Variables", Kent preprint (1973).

13. P.R. Graves-Morris and R. Hughes Jones, "An Analysis of Two Variable Rational Approximants", Brookhaven preprint AMD 674 (1974).

14. L. Hörmander, "An Introduction to Complex Analysis in Several Variables", D. van Nostrand, p.161 (1966).

THE USE OF PADE APPROXIMATION IN NUMERICAL INTEGRATION (*)

L. Wuytack

Summary.

The problem of the numerical evaluation of the integral $I = \int_a^b f(t)dt$ will be considered. A nonlinear technique, based on the use of Padé approximation, will be given and discussed.

The value of the integral satisfies $I = y(b)$, where $y(x)$ is the solution of the initial value problem $y' = f(x)$ with $y(a) = o$. This solution is approximated using Padé approximation. Consequently a one-step method is obtained to approximate the value of I. Certain properties of this method are given, e.g. its order of convergence.

1. Introduction

Most classical formulas for approximate integration of a definite integral $I = \int_a^b f(x).dx$ are linear (see [3]), which means that I is approximated by a linear combination of the values of the integrand f or

$$I \approx w_1.f(x_1) + w_2.f(x_2) + \dots + w_n.f(x_n) .$$

The points $x_1, x_2, \dots, x_n$ usually belong to $[a,b]$ and the numbers $w_1, w_2, \dots, w_n$ are called weights. In some cases the values of the derivatives of f are also taken into consideration and then linear combinations of the values of f and its derivatives at certain points are formed to approximate the value I of the integral.

(*) Work supported in part by the FKFO under grant number 2.0021.75

Using such techniques, a sequence $I_1, I_2, \ldots$ of approximate values can be constructed having I as limit. The convergence of the sequence I_k can be accelerated by several techniques. Well-known is Romberg quadrature where the sequence $\{I_k\}$ is constructed by using the trapezoidal rule and the convergence of this sequence is then accelerated by using a linear extrapolation technique. Also nonlinear acceleration techniques can be used, e.g. rational extrapolation and the ε-algorithm. A comparison of these different techniques for accelerating the convergence of $\{I_k\}$ can be found in [2] and [6] .

In many cases the linear methods for approximating I give good results. There are however situations, e.g. if f has singularities, for which linear methods are unsatisfactory. Sometimes it is then possible to modify a classical method in order to adopt it to the special situation on hand (see e.g. [4] , [7] and [8]). Another kind of approach, which will be followed in this paper, is to use a non-linear technique. This means that I will be approximated by a non-linear combination of the values of f and its derivatives at certain points.

In this paper some methods will be described which are based on the use of Padé approximation. The first method is based on the approximation of the integrand f by a Padé approximation and then performing the integration $\int_a^b r(x)dx$. This approach is called "direct" and considered in section 2. The second method is given in section 3 and is based on an "indirect" approach to the problem under consideration, by reformulating it as an initial value problem. The resulting differential equation is then solved by using Padé approximation. In order to apply this technique, the derivatives of f must be known, which might be sometimes less interesting in practice. Therefore a modification of this method is given in section 5, having the same order of convergence but without the need to compute derivatives. In section 4 some convergence properties are proved.

2. A direct method

Let R_n^m be the class of (ordinary) rational functions $r = \frac{p}{q}$ where p resp. q is a polynomial of degree at most m resp. n and such that $\frac{p}{q}$ is irreducible. Let r be the Padé approximant of order (m,n) for f in R_n^m, then $I_r = \int_a^b r(x).dx$ can be considered as an approximate value for I. In general it is however not easy to find the value of I_r, if it exists. If the poles of r are known then the partial fraction decomposition of r can be formed. This sum can then be integrated term by term in order to get I_r. This process is numerically less interesting and several difficulties can be encountered. An application of this technique to the computation of Fourier Transforms can be found in [1].

3. An indirect method

Put $y(x) = \int_a^x f(t).dt$ then $I = y(b)$. If f is Riemann integrable on $[a,b]$ then y is continuous on $[a,b]$. Furthermore if f is continuous at a point x of $[a,b]$, then y is differentiable at x and $y'(x) = f(x)$. The computation of I can now be done by computing the value in b of the solution y of the following initial value problem :

$$y'(x) = f(x) \ , \ y(a) = o \ . \qquad (1)$$

In order to find a solution of (1) numerically a discretization technique can be used. Let $x_i = a + i.h$ for $i = 0,1,\ldots;M$ with M a positive integer and $h = \frac{b-a}{M}$. Let $y_0 = y(a)$ and an approximate value of $y(x_i)$ for $i > 0$, be denoted by y_i. We now describe a method for computing y_i, for $i = 1,2,\ldots,M$.

Assume that y_i is known and that the values of the derivatives $f^{(k)}(x_i)$ exist for $k \geq 0$. Then consider the series s_i defined as follows

$$s_i(h) = y_i + h.f(x_i) + \frac{h^2}{2!}.f'(x_i) + \frac{h^3}{3!}.f''(x_i) + \ldots \quad . \tag{2}$$

Let $r_i = \frac{p_i}{q_i}$ be the Padé approximant of s_i of order (m,n). This implies that r_i is an element of R_n^m and that

$$s_i(h).q_i(h) - p_i(h) = O(h^{m+n+k_i+1}) \tag{3}$$

for some integer value of k_i, which is as high as possible.

If $q_i(x_{i+1}) \neq 0$ then the value of y_{i+1} is defined as follows

$$y_{i+1} = r_i(x_{i+1}) \quad . \tag{4}$$

Since y_0 is known and since r_i exists for every i, this technique allows us to compute $y_1, y_2, \ldots, y_M$. The value y_M can be considered as an approximate value for y(b) or I. It will be seen in the next section that $\lim_{h \to 0} y_M = I$ if certain conditions are satisfied.

A variant of the above technique has been used by P.J.S. Watson in [9], for the case where $h = b-a$ or $M = 1$. In the next example we illustrate what kind of formulas we get if (4) is used to compute approximate values for I.

Example Let $m = n = 1$ then the Padé approximant r_i of order (1,1) of (2) is the rational function associated with the irreducible form of $\frac{p}{q}$ where

$$p = y_i.f(x_i) + h.[f^2(x_i) - y_i.\frac{f'(x_i)}{2}] \quad \text{and}$$

$$q = f(x_i) - h.\frac{f'(x_i)}{2}.$$

The formula (4) gives

$$y_{i+1} = y_i + h.\frac{2.f^2(x_i)}{2.f(x_i) - h.f'(x_i)} \quad \text{for } i = 0,1,\ldots,M-1 \qquad (5)$$

It is clear that y_M is a nonlinear combination of values of f and its first derivative at the points x_i.
As a consequence of (5) we also get the following formula for approximate integration between x_i and x_{i+1} :

$$\int_{x_i}^{x_{i+1}} f(t).dt \approx h.\frac{2.f^2(x_i)}{2.f(x_i) - h.f'(x_i)}$$

4. Convergence properties

In this section some convergence properties of the method described in section 3 will be given.

Consider any one-step method of the following form

$$y_{i+1} = y_i + h.g(x_i,h) \quad \text{for } i = 0,1,\dots,M-1 \tag{6}$$

for computing a solution of (1) numerically. Assume that $g(x,h)$ is defined and continuous for every (x,h) in $[a,b] \times [0,h_o]$, where h_o is some positive real number. Under this condition the following result can be proved.

Theorem 1. Let y_i be defined by (6), then $\lim\limits_{\substack{h\to o\\ x=x_i}} y_i = y(x)$ for every x in $[a,b]$ if and only if $g(x,o) = f(x)$.

This property is a special case of a theorem about the convergence of one-step methods for the numerical solution of ordinary differential equations (see [5] ,p.71).

Due to the definition of r_i in section 3, it is clear that (4) can be written in the form (6) with $g(x,h) = [r_i(x) - y_i] /_h$ or $g(x,h) = f(x) + \frac{h}{2!} \cdot f'(x) + \frac{h^2}{3!} \cdot f''(x) + \dots$. This implies that $g(x,o) = f(x)$, consequently theorem 1 can be applied and $\lim\limits_{h\to o} y_M = y(b)$.

About the order of convergence we can prove the following result, for the case where $k_i = o$ in (3) for $i = 0,1,2,\dots,M-1$. This case is called the case of normal Padé approximants.

<u>Theorem 2.</u> Let r_i be the normal Padé approximant of s_i of order (m,n) and y_{i+1} be defined by (4), then

$$y(x_{i+1}) - y_{i+1} = O(h^{m+n+1}) \text{ as } h \to o, \text{ for } i = 0,1,2,\dots,M-1.$$

A proof of this theorem is given in [10].

5. One-step methods without using derivatives

Consider any one-step method of the form (6) for computing a solution of (1). In order to find the value of $g(x_i,h)$ it might be possible that derivatives of f must be computed. This is e.g. the case if (6) is derived by using the method in section 3. with $m+n>1$. The need to compute derivatives of the integrand can be complicated and numerically less interesting. Therefore one could try to replace $g(x_i,h)$ in (6) by another expression, without derivatives, hoping to keep a method with the same order of convergence. This technique has successfully been applied in some cases.

If e.g. the derivative in (5) is replaced by its forward difference quotient, we get

$$y_{i+1} = y_i + h \cdot \frac{2.f^2(x_i)}{3.f(x_i) - f(x_{i+1})} \tag{7}$$

It can be proved (see [10]) that this one-step method has the same order of convergence as the method defined by using (5). The formula (7) can also be considered as a nonlinear method for approximate integration of f between x_i and x_{i+1}, namely as

$$\int_{x_i}^{x_{i+1}} f(t).dt \approx h.\frac{2.f^2(x_i)}{3.f(x_i) - f(x_{i+1})}$$

6. Remarks

In [10] some numerical examples are given, illustrating the usefulness of the nonlinear techniques derived in this paper. For "smooth" integrands the classical linear methods give in general better results than the nonlinear techniques. If the integrand has a pole near the interval of integration than the nonlinear techniques can give better results. Due to possible singularities in formulas of type (5) and (7), care must be taken in applying these formulas. Difficulties can sometimes be avoided by a careful choice of the stepsize h.

Other nonlinear techniques for numerical integration are considered in [10].

Author's address

Wuytack L.
Department of Mathematics
University of Antwerp
Universiteitsplein 1
B-2610 WILRJK (Belgium)

References

[1] CHISHOLM, J.S.R.: Applications of Padé approximation to numerical integration. The Rocky Mountain Journal of Mathematics 4 (1972),159-167.

[2] CHISHOLM, J.S.R.; GENZ, A. and ROWLANDS, G.E.: Accelerated convergence of sequences of quadrature approximations. Journal of Computational Physics 10(1972),284-307.

[3] DAVIS, P.J.; RABINOWITZ, P.: Numerical integration. Blaisdell Publ. Co., London, 1967.

[4] FOX, L.: Romberg integration for a class of singular integrands. The Computer Journal 10(1967),87-93.

[5] HENRICI, P.: Discrete variable methods in ordinary differential equations. John Wiley & Sons, New York, 1962.

[6] KAHANER, D.K.: Numerical quadrature by the ε-algorithm. Mathematics of Computation 26(1972),689-693.

[7] LYNESS,J.N.; NINHAM, B.W.: Numerical quadrature and asymptotic expansions. Mathematics of Computation 21(1967),162-178.

[8] PIESSENS, R.; MERTENS, I.; BRANDERS, M.: Automatic integration of functions having algebraic end point singularities. Angewandte Informatik (1974), 65-68.

[9] WATSON, P.J.S.: Algorithms for differentiation and Integration. In "Padé approximants and their applications" (GRAVES-MORRIS, P.R., editor), 93-98. Academic Press, London,1971.

[10] WUYTACK, L.: Numerical integration by using nonlinear techniques. Journal of Computational and Applied Mathematics. To appear.

Part II

Applications of Padé Approximants Method to Problems of Fluid Mechanics

DETERMINATION OF SHOCK WAVES

BY CONVERGENCE ACCELERATION

by

Pr. Max BAUSSET - TOULON (France)

-MAY 1975-

The difficulties of the determination of stationary detached shocks arise from the global character of the problem to be solved as it is impossible to determine shock waves in the vicinity of a point. On the contrary for attached shocks determination is possible step by step from the vertex of the body.

These difficulties can be avoided by considering shock problems in a non-stationary flow. If a body situated in a motionless fluid is set into motion so that the field of initial velocities is not null, this motion immediately causes a shock wave the determination of which in the vicinity of the initial time is a local problem.

The evolution of the shocks corresponding to this motion in the vicinity of the starting point of the body is presented here. On analytical representation of the stationary detached shocks waves related to an analytical convex body can be obtained by a process of convergence acceleration, then by passing on to the limit when the permanent motion is reached. The data will be supposed to be such that operations can be considered as possible within the fields where it is being operated.

- Start of a body in a motionless fluid :

The space is related to orthonormated fixed axes.
The equation of the body is :

(1) $\quad x = f(yz) + \xi(t)$

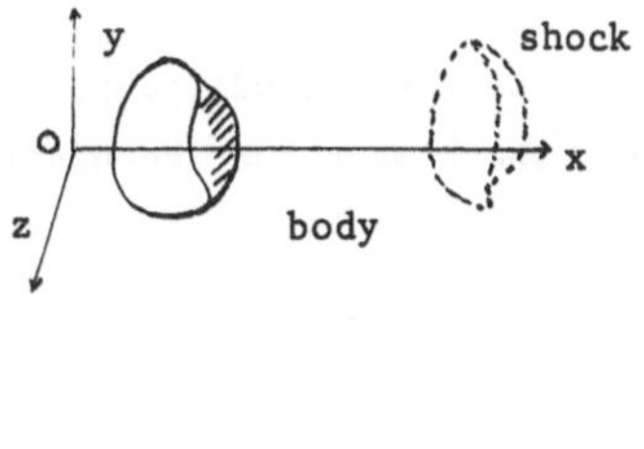

in which f is supposed to be uniform

and the body occupies the region : $x \leq f(y\ z) + \xi(t)$.

The initial data are such that :

(2) $\xi(0) = 0 \qquad \xi'(0) > 0$

This body is plunged into a non viscous compressible, non heat-conducting and supposedly perfect fluid. This motionless is defined by quantities $\bar{p}$, $\bar{\rho}$ and γ representing respectively the pressure, the density and ratio of specific heats. The latter will be supposed to be constant throughout the motion. The characteristic quantities of this fluid (velocity, pressure and density) will be designated as V, p and ρ at the point of the spatio-temporal coordinates x, y, z, and t.

The fluid and the body are motionless before t = 0. Since at the initial time $\xi'(0) > 0$ the motion of the body immediately causes a schock wave which propagates itself through the fluid. Its equation will be :

(3) $x = F(y\ z\ t)$

For reasons of calculations symmetry which will appear below, let us write :

(4) $\phi(x\ y\ z\ t) \equiv f(y\ z) + \xi(t) - x$

$\Phi(x\ y\ z\ t) \equiv F(y\ z\ t) - x$

so that when considering covector $\bar{r} = |\ x, y, z\ |$ the vectors normal at each instant to any point of the body and the shock are defined by :

(5) $N_\phi = \overline{\delta_r\ \phi} = \begin{vmatrix} -1 \\ f'_y \\ f'_z \end{vmatrix} \qquad N_\Phi = \delta_r\ \Phi = |-1, F'_y, F'_z|$

In these conditions, the normal number of Mach M at time t at any point of the shock wave and the upstream number of Mach at infinite $\mathcal{M}$ are defined by :

(6) $$M(\Phi) = \frac{\delta_t \Phi}{c(\delta_r \Phi \times \delta_r \Phi)^{\frac{1}{2}}} \qquad \mathcal{M} = -\frac{\xi'(0)}{\bar{c}}$$

c designating the speed of sound so that $c^2 \rho = \gamma p$ at any point x, y, z, t. Let us write $(\gamma + 1)\mu^2 = \gamma - 1$ and relate pressure and density to their valours at infinite introducing without dimensions :

$$P(x\ y\ z\ t) = \frac{p}{\bar{p}} \qquad R(x\ y\ z\ t) = \frac{\bar{\rho}}{\rho}$$

In these conditions, the continuous motions of the considered fluid are defined by the classic equations of dynamics, mass and energy conservations which will be written under the form :

(7)
$$d_t V + \bar{c}^2 \frac{1-\mu^2}{1+\mu^2} . R . \overline{\delta_r P} = 0$$
$$d_t R - R .. T_r (\delta_r V) = 0$$
$$(1-\mu^2) R . d_t P + (1+\mu^2) P . d_t R = 0$$

d_t designating the corpuscular derivative and Tr the trace of matrix $\delta_r V$.

The solution of shock problems is a result of the study of this system and the boundary of limits on the body and on the shock wave.

In the absence of viscosity, the body is necessarily a stream surface, which leads to the equations :

(8)
$$\phi(x\ y\ z\ t) = 0$$
$$\delta_r \phi \times V_\phi = \delta_t \phi$$

in which V_ϕ is the value of V on the body $\phi = 0$.

The classic conditions of shock phenomena mean that the quantities of motion, mass and energy are preserved while crossing the surface of the wave. These usual conditions can be expressed by the equations :

$$\Phi\,(x\ y\ z\ t\,) = 0$$

(9)

$$V_\Phi = -\,(1-\mu^2)\cdot\left\{\frac{\delta_t\Phi}{\delta_r\Phi \times \overline{\delta_r\Phi}} - \frac{\bar{c}^{\,2}}{\delta_t\Phi}\right\}\cdot\ \overline{\delta_r\Phi}$$

$$P_\Phi = \frac{1+\mu^2}{\bar{c}^2}\cdot\left\{\frac{\delta_r\Phi \times \overline{\delta_r\Phi}}{(\delta_t\Phi)^2}\right\}^{-1} - \mu^2$$

$$R_\Phi = \bar{c}^2(1-\mu^2)\cdot\left\{\frac{\delta_r\Phi \times \overline{\delta_r\Phi}}{(\delta_t\,\Phi)^2}\right\} + \mu^2$$

in which V_Φ , P_Φ and R_Φ are the values of V, P and R on shock $\Phi = 0$.

At any point of the spatio-temporal region included between the body and the shock and for $t \geq 0$ equations (7) are identities in x, y, z and t. Added to equations (8) and (9), they permit the calculation of the partial n-order derivatives of quantities V, P, R and Φ at time $t = 0$.

- Initial velocity of a non-stationary shock :

In effect, if one place oneself at the initial time when the fluid is motionless everythere except on the body which is set into motion, the position of the shock wave coincides with the position of the body. Thus one has for $t = 0$ the relations :

(10) $\Phi\,(x\ y\ z\ 0) = \phi\,(x\ y\ z\ 0)$ $\qquad F(y\ z\ 0) = f\,(y\ z)$

which will be designated $\Phi_o = \phi_o$.

They entail the following equations :

(11)

$$V_{\Phi_o} = V_{\phi_o} \qquad R_{\Phi_o} = R_{\phi_o}$$

$$P_{\Phi_o} = P_{\phi_o} \qquad \delta_r\,\Phi_o = \delta_r\,\phi_o$$

From the given equations (8) and (9) considered at time t = 0 one deduces the following relations :

(12)
$$\delta_r\phi_o \times V_{\phi_o} = (\delta_t\phi)_{t=0} = \xi'(0)$$

$$\delta_r\Phi_o \times V_{\Phi_o} = -(1-\mu^2)\left(\frac{(\delta_t\Phi)_{t=0}}{\delta_r\Phi_o \times \overline{\delta_r\Phi_o}} - \frac{\bar{c}^2}{(\delta_t\Phi)_{t=0}}\right)(\delta_r\Phi_o \times \overline{\delta_r\Phi_o})$$

If one notices that $(\delta_t\Phi)_{t=0} = F'_t\,(y\ z\ 0)$ one deduces by elimination of V the initial value of the shock velocity at any point :

(13)
$$F'_t\,(y\ z\ 0) = \frac{\xi'(0)}{2(1-\mu^2)} + \left[\left(\frac{\xi'\,(0)}{2(1-\mu^2)}\right)^2 + \delta_r\phi_o \times \overline{\delta_r\phi_o}\right]^{\frac{1}{2}}$$

- Second order derivatives :

The total number of the partial n-order derivatives of a function A(x y z t) being C^n_{n+3}, one will be brought to consider the table :

(14)
$$\delta^n_{r^n}\ A\,(x\,y\,z\,t) = \left\| \frac{\delta^n A}{\delta x^n}\ ;\ \frac{\delta^n A}{\delta y^n}\ ;\ \frac{\delta^n A}{\delta z^n} \right\|$$

which will be a line or a matrix 3 x 3 according as A has a scalary or vectorial value.

Relations (11) then permit to calculate the spatial derivatives of (x y z t) for t = 0 in the form :

(15)
$$\delta^n_{r^n}\,(\delta_r\phi_o) = \left\| \frac{\delta^n}{\delta x^n}(\delta_r\phi_o)\ ;\ \frac{\delta^n}{\delta_y{}^n}(\delta_r\phi_o)\ ;\ \frac{\delta^n}{\delta_z{}^n}(\delta_r\phi_o) \right\|$$

from which the initial values of all the spatial partial derivatives of the function representing the shock wave :

(16)
$$\frac{\delta^{p+q}\ F\,(y\ z\ 0)}{\delta y^p\ \ \delta z^q} = \frac{\delta^{p+q}\ f\,(y\ z)}{\delta y^p\ \ \delta z^q}$$

can be deduced.

But concerning the temporal or spatio temporal derivatives of F it is necessary to use the derivatives of the boundary of limits on the body and the shock.

If one designates by δ one differential which according to the case can be δ_r, δ_t, or d_t, one deduces from equations (8) and (9) the following linear equations in relation to $\delta(\delta_t\phi)$, $\delta(\delta_t\Phi)$ and $\delta(\delta_r\Phi)$:

(17)
$$\phi(x\ y\ z\ t) = 0$$
$$\delta_r\phi \times \delta V_\phi = \delta(\delta_t\phi) - \delta(\delta_r\phi) \times V_\phi$$

(18)
$$\Phi\ (x\ y\ z\ t) = 0$$
$$\delta\ V_\Phi = \alpha.\ \overline{\delta_r\Phi} \times \delta(\delta_t\Phi) + \beta.\ \overline{\delta_r\Phi} \times \delta_r\Phi\ .\ \overline{\delta(\delta_r\Phi)} + \gamma\ .\ \overline{\delta(\delta_r\Phi)}$$
$$\delta\ P_\Phi = \alpha_1.\ \delta(\delta_t\Phi) + \beta_1.\ \overline{\delta(\delta_r\Phi)}$$
$$\delta\ R_\Phi = \alpha_2.\ \delta(\delta_t\Phi) + \beta_2.\ \overline{\delta(\delta_r\Phi)}$$

in which to following quantities which are dependent only on the first order derivatives of Φ that are known at the initial time are :

$$\alpha = -(1-\mu^2).\frac{\bar{c}^2\,\delta_r\Phi \times \overline{\delta_r\Phi} + (\delta_t\Phi)^2}{\delta_r\Phi \times \overline{\delta_r\Phi}\,.(\delta_t\Phi)^2} \qquad \beta = 2(1-\mu^2).\frac{\delta_t\Phi}{\delta_r\Phi \times \overline{\delta_r\Phi}}$$

(19)

$$\gamma = (1-\mu^2).\frac{\bar{c}^2\,\delta_r\Phi \times \overline{\delta_r\Phi} - (\delta_t\Phi)^2}{\delta_r\Phi \times \overline{\delta_r\Phi}\,.(\delta_t\Phi)^2} \qquad \alpha_1 = \frac{2(1+\mu^2)}{\bar{c}^2}.\frac{\delta_t\Phi}{(\delta_r\Phi \times \overline{\delta_r\Phi})^2}$$

$$\overline{\beta_1} = -\frac{2(1+\mu^2)}{\bar{c}^2}.\frac{(\delta_t\Phi)^2}{(\delta_r\Phi \times \overline{\delta_r\Phi})^2}\,.\,\overline{\delta_r\Phi} \qquad \alpha_2 = -2\,\bar{c}^{-2}(1-\mu^2)\frac{\delta_r\Phi \times \overline{\delta_r\Phi}}{(\delta_t\Phi)^4}$$

$$\overline{\beta_2} = 2\,\bar{c}^{-2}.(1-\mu^2).\frac{\overline{\delta_r\Phi}}{(\delta_t\Phi)^2}$$

According to the choice of differential δ, equations (18) include the scaleries, lines and matrices :

$$\delta_t(\delta_t\Phi) = \Phi''_{t^2} \qquad \delta_t(\delta_r\Phi) = \left| \Phi''_{tx} \; \Phi''_{ty} \; \Phi''_{tz} \right|$$

(20)
$$\overline{\delta_r(\delta_r\Phi)} = \begin{vmatrix} \Phi''_{x^2} & \Phi''_{xy} & \Phi''_{xz} \\ \Phi''_{yx} & \Phi''_{y^2} & \Phi''_{yz} \\ \Phi''_{zx} & \Phi''_{zy} & \Phi''_{z^2} \end{vmatrix} \qquad \delta_r(\delta_t\Phi) = \delta_t(\delta_r\Phi)$$

which contain all the partial second order derivatives of function Φ. As to the first members, they contain the derivatives of $\mathbf{V}$, P and R on the body or the shock, i.e. the quantities ;

$$d_t(V_\Phi) = \frac{\delta V_\Phi}{\delta x}\,\phi'_t + \frac{\delta V_\Phi}{\delta t}$$

for the scalery derivative and the tables :

$$\delta_r(V_\Phi) = \left| \frac{\delta V_\Phi}{\delta x}\,\Phi'_x \; ; \; \frac{\delta V_\Phi}{\delta y} \; ; \; \frac{\delta V_\Phi}{\delta z} \right|$$

for the vectorial derivative and the similar quantities for P and R and for function Φ.

Considering the equations (7) of the motion valid on the body $\phi = 0$ with the preceding differential relations, one has eleven sixteen indeterminate equations. Namely δV, δP, δR, on the shock or on the body as well as $\delta(\delta_t\Phi)$ and $\delta(\delta_r\Phi)$. Consequently the process used does not permit, as in the case of first derivatives, to calculate all the second order derivatives of V, P, R and F placing oneself at any time.

- Initial acceleration of a non-stationary shock :

If one places oneself at time t = 0 the preceding relations entail :

(21)
$$d_t(V_{\phi_o}) = \frac{\delta V_{\phi_o}}{\delta x}\,\xi'(0) + \frac{\delta V_{\phi_o}}{\delta t}$$
$$d_t(V_{\Phi_o}) = \frac{\delta V_{\Phi_o}}{\delta x}\,F'_t(0) + \frac{\delta V_{\Phi_o}}{\delta t}$$

for corpuscular derivatives as well as equations :

$$\left\| -\frac{\delta V_{\phi_o}}{\delta x} \; ; \; \frac{\delta V_{\phi_o}}{\delta y} \; ; \; \frac{\delta V_{\phi_o}}{\delta z} \right\| = \left\| -\frac{\delta V_{\Phi_o}}{\delta x} \; ; \; \frac{\delta V_{\Phi_o}}{\delta y} \; ; \; \frac{\delta V_{\Phi_o}}{\delta z} \right\|$$

(22) $$\frac{\delta V_{\phi_o}}{\delta t} = \frac{\delta V_{\Phi_o}}{\delta t}$$

and the following value :

(23) $$\delta_r(\delta_r \Phi_o) = \begin{vmatrix} 0 & 0 & 0 \\ 0 & f''^2_y & f''_{yz} \\ 0 & f''_{zy} & f''^2_z \end{vmatrix} = \delta_r\,(\delta_r f)$$

One sees then that at that time the ensemble of second order partial derivatives is determined.

In particular, one obtains :

$$\delta_t(\delta_t \Phi_o) = F''_{t^2}(y\ z\ 0) \qquad \delta_{rt}(\Phi_o) = \left| 0 \; ; \; F''_{ty}(y\ z\ 0) ; \; F''_{tz}(y\ z\ 0) \right|$$

If one introduces the vector normal to the body for $t = 0$: $\overline{N_{\phi_o}} = \delta_r f$ and the normal initial number of Mach defined from (6) by :

(24) $$M\ (x\ y\ z\ 0) = \mathcal{M}.\left(\delta_r f \times \overline{\delta_r f}\right)^{-\frac{1}{2}}$$

the results are as follows :

(25) $$\left| 0 ; F''_{ty}(y\ z\ 0) ; F''_{tz}(y\ z\ 0) \right| = \frac{2\,M\,\overline{c}}{M^2+1} \cdot \left(\delta_r f \times \overline{\delta_r f}\right)^{-\frac{1}{2}} . \left| 0 \; ; \; \overline{N}_{\phi_o} \frac{\delta N_{\phi_o}}{\delta y} \; ; \; \overline{N}_{\phi_o} \frac{\delta N_{\phi_o}}{\delta z} \right|$$

for all initial spatio-temporal derivatives.

The initial value of shock acceleration is expressed as follows :

(26) $$F''_{t^2}\ (y\ z\ 0) = \frac{M^2(M^2-1)}{(1-\mu^2)C_4} \cdot \xi''(0) - \overline{c}^2 \frac{A_6}{C_4} . \mathcal{L}_1(f) + \overline{c}^2 \frac{A_8}{C_6} . \mathcal{L}_2(f)$$

in which A_n and C_n are n-degree polynomial expressions and $\mathcal{L}_1(f)$ and $\mathcal{L}_2(f)$ quantities related to the geometry of the body :

(27)

$$A_6 = (\mu^2 M^2 + 1 - \mu^2)\cdot(M^2-1) \left[(1 + \mu^2) M^2 - \mu^2\right]$$

$$A_8 = (\mu^4 - \mu^2) M^8 + (-2\mu^4 + 3\mu^2 + 1) M^6 + (3\mu^4 + 2\mu^2 -1) M^4 + +(-2\mu^4 + \mu^2+1)M^2 \; -(1- \mu^2)^2$$

$$C_4 = (3 \mu^2 + 1) M^4 - (2\mu^2 - 3) M^2 - \mu^2$$

$$C_6 = (M^2 + 1) \left[(3 \mu^2 + 1) M^4 - (2\mu^2 -3) M^2 - \mu^2\right]$$

(28)

$$\mathcal{L}_1 (f) = 2\, \delta_r f \times \delta_r(\delta_r f) \times \overline{\delta_r f}.\ \langle\delta_r f \ \times \overline{\delta_r f}\rangle^{-1} - \mathrm{Tr}\ \langle\delta_r(\delta_r f)\rangle$$

$$\mathcal{L}_2 (f) = \delta_r f \times \delta_r(\delta_r f) \times \overline{\delta_r f}.\ \langle\delta_r f \times \overline{\delta_r f}\rangle^{-1}$$

If one places oneself in the particular case of plane or axisymmetrical flows one finds again in (13) and (26) formulae established by CABANNES.

- <u>About superior order derivatives</u> :

The process used for the calculation of initial second order derivatives can be applied to superior order derivatives.

The total number of the n-order derivatives of wave function F(y z t) is C^n_{n+2}. Now if one differentiates identities (17) and (18) in y, z, and t (n-1) times and then if one places oneself at time t = 0 one obtains 6 C^{n-1}_{n+1} identities in relation to y and z.

Similarly if one derives (n-2) times the equations of the motion which are identities in relation to x, y, z and t one obtains 5 C^{n-2}_{n+1} relations.

Placing oneself again at time t = 0 and on the shock or the body one finally obtains :

(29) $$f(n) = 6\, C^{n-1}_{n+1} + 5\, C^{n-2}_{n+1}$$

equations. These are linear in relation to the initial partial derivatives of V p and ρ numbering 5 C^{n-1}_{n+2} and to the initial values of the (n-1) order derivatives of F(y z t) numbering C^{n-1}_{n+1}. Indeed, the coefficients of these equations are dependent only on derivatives the order of which is inferior to (n-1).

Consequently, giving to n the integer values successive from n = 2 one can calculate all the preceding indeterminates as one can see that the number of equations equales the number of indeterminates.

The total number of the equations to be resolved is therefore :

(30) $$N = \sum_{2}^{n} f(n) = -6 + \frac{n(n+1)(n+2)(5n+19)}{4!}$$

In particular, one can thus calculate derivatives $F^{(n)}_{tn}(y\ z\ 0)$. This calculation including all the initial scalery derivatives of the three characteristic functions V, p and ρ can be simplified as it has been shown in the case of second order derivatives considering the lines :

(31) $$\delta^{(n)}_{t^n}(\delta_r\Phi) = \left| \Phi^{(n+1)}_{t^n x} \; ; \; \Phi^{(n+1)}_{t^n y} \; ; \; \Phi^{(n+1)}_{t^n z} \right|$$

as well as tables (14) for A = V, p, ρ in the form :

(32) $$\delta^{(p+q)}_{r\ ptq} A(x\ y\ z\ t) = \left\| \frac{\delta^{(p+q)} A}{\delta x^p \delta t^q} \; ; \; \frac{\delta^{(p+q)} A}{\delta y^p \delta t^q} \; ; \; \frac{\delta^{(p+q)} A}{\delta z^p \delta t^q} \right\|$$

One can see that one has thus to resolve according to the values of n successively :

n	2	3	4	5	6	....
f(n)	23	56	110	150	301	....
N	23	79	189	379	680	....

scalery equations the number of which can be reduced thanks to (31) and (32).

In particular A, B ..., H, ... designating polynominal expressions in relation to M the degree of which is equal to the index, one obtains :

$$F'''_{t^3}(y\ z\ 0) = \frac{M^2}{6(1-\mu^2)(M^2+1)} \cdot \xi'''(0) + \bar{c} \cdot \frac{A_8}{B_8} \cdot \xi''(0) \cdot \frac{\mathcal{L}_1(f)}{(\mathcal{L}_2(f))^2} +$$

(33) $$+ \frac{1}{\bar{c}} \frac{B_{10}}{C_{12}} \cdot \xi''(0) \cdot \mathcal{M}_1(f) + \bar{c}^3 \frac{D_{12}}{E_{12}} \cdot \mathcal{M}_2(f) + \bar{c}^3 \cdot \frac{F_{12}}{G_{12}} \cdot \mathcal{M}_3(f) +$$

$$+ \bar{c}^3 \frac{F_{14}}{E_{16}} \cdot \mathcal{M}_4(f) + \bar{c}^3 \cdot \frac{D_{14}}{C_{16}} \cdot \mathcal{M}_5(f) + \bar{c}^3 \cdot \frac{H_{16}}{L_{16}} \cdot \mathcal{M}_6(f)$$

in which $\mathcal{M}_i(f)$ are still quantities dependent on the body. One obtains :

(34) $$\mathcal{M}_3(f) = \left\langle \delta_r f \times \delta^{(2)}_{r^2}(\delta_r f) \times \overline{\delta_r f} \right\rangle^2 \frac{1}{\left\langle \delta_r f \times \overline{\delta_r f} \right\rangle^{3/2}} +$$
$$+ \operatorname{Tr} \left\langle \delta^{(2)}_{r^2}(\delta_r f) \right\rangle - \det \left\langle \delta_r(\delta_r f) \right\rangle \cdot \left\langle \delta_r f \times \delta_r(\delta_r f) \times \overline{\delta_r f} \right\rangle$$

$F^{(IV)}_{t^4}(y\ z\ 0)$ given the considerable complexity of calculations ; it has been impossible to develop $F^{(IV)}_{t^4}$ except for $y = z = 0$ corresponding to the apex of the shock wave.

- Approximate equations of non-stationary shocks :

The proceding calculations permit to write the equation of non-sta stationary shocks in the form of a 4-degree polynominal expression :

$$F\ (y\ z\ t) = f\ (y\ z) + t\ F'_t\ (y\ z\ 0) + \frac{t^2}{2} F''_{t^2}(y\ z\ 0) + \tag{35}$$

$$+ \frac{t^3}{6} F'''_{t^3}(y\ z\ 0) + \frac{t^4}{24} F^{IV}_{t^4}\ (y\ z\ 0)$$

This represents with tolerable approximation these shock waves for three spatial variable flows and its evolution in the vicinity of the starting point of the body. Now the considered shock waves exist whatever value t has as long as velocity remains supersonic. It is therefore interesting to approach the equation of these waves by an analytical expression valid within the longest possible lapse of time.

If function F (y z t) is uniform, one can accelerate the convergence of the associated Taylor series. WYNN'S ε-algorithm corresponding to PADE'S diagonal approximation will be chosen. Thus the fractionary approximation of the equation of non-stationary shocks will be determined in the form :

$$[1,1]\ F\ (yzt) = \frac{\left[f(yz)\ F''_{t^2}(y\ z\ 0) - 2\ F'_t\ (y\ z\ 0)\right]\ t - 2f(yz).F'_t(y\ z\ 0)}{t.\ F''_{t^2}\ (y\ z\ 0) - 2\ F'_t\ (y\ z\ 0)} \tag{36}$$

whose field of definition is that of PADE'S first order diagonal approximations.

It is interesting to notice that in the case of very high velocities of the body $\mathcal{M} \to \infty$ one can write :

$$M\ (y\ z\ 0)\ \sim\ \mathcal{M}(\delta_r f \times \overline{\delta_r f})^{-\frac{1}{2}} \qquad F'_t\ (y\ z\ 0)\ \sim\ \frac{\mathcal{M}}{1-\mu^2}$$

$$F''_{t^2}(y\ z\ 0)\ \sim\ \frac{-\mu^2\ \bar{c}^2}{3\mu^2+1}\left[(1+\mu^2).\mathcal{L}_1(f) + (1-\mu^2).\mathcal{L}_2(f)\right].\mathcal{M}^2$$

One can then see that, for an analytical body, approximation (36) remains valid for $t > 0$. It is clear that such is not the case if the body's velocity is indeterminate.

- Détermination of detached stationary shocks :

It is possible to infer from the foregoing results approximate formular determining the positions of the detached stationary shocks created by the movements of blunt analytical bodies in translation motion.

In effect, when the second derivative $\xi''(t)$ vanishes for $t \to \infty$ the motion of the body tends to become uniform.
It is then logical to admit that the motion of the fluid in relation to a mark linked to the body tends to become stationary. Consequently the position of a stationary shock results from the knowledge of F(y z t) for $t \to \infty$. It is similary logical to admit that the stationary flow is reached all the more rapidly as the velocity limit of the body is reached more rapidly. One will place oneself in the case when is chosen so that :

$$t \leqslant 0 : \xi (t) = 0 \qquad\qquad t > : \xi (t) = t . \xi' (0)$$

In any space of time when F(y z t) is an analytical variable dependent on time, one can write :

$$(37) \qquad F (y\ z\ t) = \sum_{j=0}^{\infty} a_j(yz) . t^j$$

and one has indicated a calculation process of a_j coefficients. There are several methods allowing to replace the analytical function $\sum_0^\infty a_j t^j$ by a sequence of functions converging towards F(yz) in the field of convergence of the entire sequence $a_j t^j$ and which are definite whatever the value of t is and which have a limit when time increases indefinitely.

Within the framework of the ε-algorithm theory supposes

$$(38) \qquad A_n = \sum_0^n a_j t^j \qquad\qquad n > 0$$

is a converging sequence whose terms are dependent as parameters.
Among others this theory proposes to replace this sequence by another converging more rapidly towards the same limit or converging to a more extensive field than the initial sequence.

PADE'S n-order diagonal approximation (used in (36) for n = 1) is defined by :

$$(39)\qquad [n,\, n]\, A_n = \frac{\begin{vmatrix} t^n A_0 & ; & t^{n-1} A_1 ; \cdots\cdots & t^{\circ} A_n \\ a_1 & ; & a_2 ; \cdots\cdots & a_{n+1} \\ \vdots & & & \vdots \\ a_n & ; & a_{n+1} ; \cdots\cdots & a_{2n} \end{vmatrix}}{\begin{vmatrix} t^n & ; & t^{n-1} ; \cdots\cdots & t^{\circ} \\ a_1 & ; & a_2 ; \cdots\cdots & a_{n+1} \\ \vdots & & & \vdots \\ a_n & ; & a_{n+1} ; \cdots\cdots & a_{2n} \end{vmatrix}} = \frac{P_n(t)}{Q_n(t)}$$

which leads in particular to :

$$\lim_{t\ \infty} \frac{P_1(t)}{Q_1(t)} = \frac{a_0 a_2 - a_1^2}{a_2}$$

$$(40)\qquad \lim_{t\ \infty} \frac{P_2(t)}{Q_2(t)} = \frac{a_0(a_1a_4 - a_3^2) - a_1(a_1a_4 - a_2\, a_3) + a_2(a_1a_3 - a_2^2)}{a_2a_4 - a_3^2}$$

$$\lim_{t\ \infty} \frac{P_n(t)}{Q_n(t)} = (-1)^n \frac{\Delta}{\delta}$$

with :

$$\Delta = \begin{vmatrix} a_0 & , & a_1 , \cdots\cdots & a_n \\ a_2 & , & a_3 , \cdots\cdots & a_{n+1} \\ \vdots & & & \vdots \\ a_n & , & a_{n+1} , \cdots\cdots & a_{2n} \end{vmatrix}$$

and δ being the cofactor of a_0 in Δ.

As to SHANKS'S transformations, they are defined by associating to sequence A_n the sequence :

$$(41)\qquad B_n = \frac{A_{n+1} \cdot A_{n-1} - A_n^2}{A_{n+1} + A_{n-1} - 2\, A_n} \qquad n \geq 1$$

The process can be iterated by considering the sequence :

$$(42)\qquad C_n = \frac{B_{n+1}\,B_{n-1} - B_n^2}{B_{n+1} + B_{n-1} - 2B_n} \qquad n \geq 2$$

and so forth.

In each of these iterations the first term $B_1\ C_2$ ... has a limit for $t \to \infty$

$$\lim\ B_1(t) = \frac{a_o a_2 - a_1^2}{a_2}$$

$$(43)\qquad \lim\ C_2(t) = \frac{a_o a_2 - a_1^2}{a_2} - \frac{(a_1 a_3 - a_2^2)^2}{a_4 a_2 - a_3^2} \cdot \frac{a_4}{a_2^2}$$

One can then see that PADE'S approximation $[1,1]$ and SHANKS'S first order B_1 have the same limit which we shall call first order approximation. On the contrary, second order approximation differ.

- Fractionary approximation of stationary shocks :

When the studied function is determinate, which is the case here, the choice between an approximation method and another can only be made through a comparison with accurate numerical results relative to know particular cases.

Concerning the approximation of function $x = F(y\ z\ t)$ representing the stationary detached wave, the preceding calculations have been conducted to an indefinite point (x, y z) only up to n = 2.

Placing oneself at a mark linked to the body, it follows from (35) and :

$$(44)\qquad F'_t\ (y\ z\ 0) - \xi'(0) = \bar{c}\cdot\frac{\mu^2 M^2 + 1 - \mu^2}{M}\cdot\left(\delta_r f \times \overline{\delta_r f}\right)^{\frac{1}{2}}$$

that in the vicinity of the initial time one can write :

$$(45)\qquad F\ (y\ z\ t) = f\ (y\ z) + \bar{c}t\ \frac{\mu^2\ M^2 + 1 - \mu^2}{M}\ \left(\delta_r f \times \overline{\delta_r f}\right)^{\frac{1}{2}} + \frac{t^2}{2} F''_{t^2}\ (y\ z\ 0)$$

in which the expression of the second derivative follows from (26) with $\xi''(0) = 0$. In these conditions PADE'S (or SHANKS'S) first order approximation gives a position of the detached stationary shock :

(46) $$x = f(y\ z) - \frac{2(\mu^2 M^2 + 1 - \mu^2)^2 . (M^2 + 1) . C_4}{M^2 . A_8 . \mathcal{L}_2(f) - M^2 . (M^2+1) . A_6 . \mathcal{L}_1(f)} . (\delta_r f \times \overline{\delta_r f})$$

In this expression the velocity of the body and the conditions of the motionless fluid interven through the normal number of Mach which plays the role of an auxiliary parameter. There results a certain complication even for very simple analytical bodies.

As shown above the permanent motion will be reached all the more rapidly as the body's velocity is high. In this hypothesis the preceding relation is consideratly simplified and leads to the approximation :

(47) $$x = f\ (y\ z) + \frac{2\ \mu^2(3\ \mu^2 + 1)}{(1 + \mu^2) . \mathcal{L}_1(f) + (1 - \mu^2) . \mathcal{L}_2(f)} \cdot \{\delta_r f \times \overline{\delta_r f}\} + \mathcal{O}\left(\frac{1}{\mathcal{M}^2}\right)$$

The field of validity of (46) remains to be specified in relation to $\mathcal{M}$ as well as to r. As in the case of (47) it is linked to that of PADE'S approximations for the accelerated sequence.

It can easily be verified that if analytical function f(y z) is chosen as uniform and convex everywhere, the field of definition of (47) remains limited.

The application of (46) or (47) to the two dimensional flows (or revolutions) leads for the body's curve $x = \phi\ (y)$ to

(48) $$\Phi(y) = \phi(y) - \frac{2\ \mu^2(3\ \mu^2 + 1)(1 + \phi'^2)^2 y}{(1 + \mu^2) y \cdot \phi'' - 2y \cdot \phi'' . \phi'^2 + \varepsilon\ (1 + \mu^2) . (1 + \phi'^2)\phi'} + \mathcal{O}\left(\frac{1}{\mathcal{M}^2}\right)$$

and :

(49) $$\Phi(y) = \phi(y) - \frac{2\ (\mu^2 M^2 + 1 - \mu^2)^2 . (M^2 + 1) . C_4 (1 + \phi'^2) \cdot y}{\left\{ \begin{array}{l} [M^2 A_8 - 2(M^2 + 1) A_6] \cdot y \cdot \phi'^2 \phi'' + M^2(M^2 + 1) A_6 (1 + \phi'^2) y \phi'' \\ + \varepsilon\ M^2\ (M^2 + 1)\ A_6\ (1 + \phi'^2) \cdot \phi \end{array} \right\}}$$

with $M = \mathcal{M}(1 + \phi'^2)^{\frac{1}{2}}$

coefficient ε being equal to 0 or 1 wether the problem is 2-dimensional or axisymmetrical in relation to Ox.

- Distance of vertices :

In order further to specify relation (46), suppose that the body has been chosen so that for y = z = 0 one has : $\delta_r f = |-1, 0, 0|$.
The corresponding point will be called vertex of the body. Designating R_1 and R_2 the principal radii of curvature at this point, one has with the utilized notations :

$$(50) \qquad \begin{aligned} \omega^{\cdot} &= \frac{1}{R_1} + \frac{1}{R_2} = \mathrm{Tr}\ \langle\delta_r(\delta_r f)\rangle \\ \Pi &= \frac{1}{R_1\ R_2} = \mathrm{Det}\ \langle\delta_r(\delta_r f)\rangle \end{aligned}$$

The vertex of the shock surface will be the point of the wave situated on axis Ox.

The calculations have been conducted at the vertex up to n = 4. One can therefore work out not only PADE'S approximation [1,1] but also SHANKS'S second order approximation ⟨formulae (43)⟩.

Designating h_1 the first order approximation of the distance of vertices one obtains :

$$(51) \qquad h_1 = \frac{1}{\omega} \cdot \frac{2(\mu^2 M_o^2 + 1 - \mu^2)\cdot\left[(3\mu^2+1)\ M_o^4 - (2\mu^2-3)M_o^2 - \mu^2\right]}{M_o^2\cdot\left[(1+\mu^2)\ M_o^2 - \mu^2\right]\cdot(M_o^2-1)}$$

In the second order approximation considered here one has :

$$(52) \qquad h_2 = h_1 \cdot \left[1 + \frac{3}{16}\cdot\frac{P_4\cdot S_8}{Q_4\cdot T_8}\right]$$

In which one has :

$$(53) \qquad \begin{aligned} P_4 &= M_o^4 - (4\mu^2-1)\ M_o^2 - 4\ (1-\mu^2) \\ Q_4 &= M_o^4 + (2\mu^2+1)\ M_o^2 + 2\ (1-\mu^2) \end{aligned}$$

$$\begin{aligned} S_8 = M_o^8 &+ (-6\mu^4 + 8\mu^2 + 1)\ M_o^6 + (18\mu^4 - 12\mu^2 + 9)\ M_o^4 + \\ &+(-18\ \mu^4 + 16\ \mu^2 + 2)\ M_o^2 + 6\ (I - \mu^2)^2 \end{aligned}$$

$$T_8 = M_o^8 + (18\,\mu^4 - 8\,\mu^2 + 2)\,M_o^6 - (38\mu^4 - 36\,\mu^2 + 7)\,M_o^4 + \tag{53}$$
$$+(22\,\mu^4 - 17\mu^2 + 10)\,M_o^2 - 2\,(1-\mu^2)^2$$

and in which M_o is the normal initial number of Mach at the vertex :

$$M_o = \frac{\mathcal{M}}{2(1-\mu^2)} + \left[\frac{\mathcal{M}^2}{4(1-\mu^2)^2} + 1\right]^{\frac{1}{2}} \tag{54}$$

It may seem paradoxical that the second order approximation is only dependent on $\delta_r(\delta_r f)$ and not on third order derivatives. This follows from the fact that one is placed at point $y = z = 0$

Again, placing oneself in the field of high velocities the preceding expressions had to the approximate formulae :

$$h_1 = \frac{1}{\omega} \cdot \frac{2\mu^2(3\mu^2+1)}{1+\mu^2} + \mathcal{O}\left(\frac{1}{\mathcal{M}^2}\right) \tag{54}$$
$$h_2 = h_1 \cdot (1 + \frac{3}{16}) + \mathcal{O}\left(\frac{1}{\mathcal{M}^2}\right)$$

- <u>Curvatures of shocks at the vertex</u> :

In order to calculate the principal radii of curvature ρ_1 and ρ_2 at the point of the shock wave situated on Ox, it is necessary to calculate the spatial second order derivatives of F(y z t). Now, these derivatives are known only up to $n = 2$ when one derives also in relation to time. It is therefore only possible to apply PADE'S [1, 1] transformation.

$$\frac{1}{\rho_1} + \frac{1}{\rho_2} = \omega - \frac{(\omega^2 - 2\pi)^2 \cdot R_6}{\omega^3 \cdot S_8 - 2\,\pi\omega \cdot T_8} \tag{55}$$
$$\frac{1}{\rho_1\rho_2} = \pi - \frac{\omega^2\,\pi \cdot R_6}{\omega^2 \cdot S_8 - \pi \cdot R_6}$$

with :

$$R_6 = 4\, M_o^2 \left[(3\mu^2 + 1)\, M_o^4 - (2\mu^2 - 3)\, M_o^2 - \mu^2\right]$$

(56)

$$S_8 = -(\mu^4 - 6\mu^2 - 3)\, M_o^8 + (4\mu^4 + 3\mu^2 + 7)\, M_o^6 + (-6\mu^4 - 7\mu^2 + 7)\, M_o^4 + \\ +(4\mu^4 - 3\mu^2 - 1)\, M_o^2 - \mu^4 + \mu^2$$

$$T_8 = -(\mu^4 - 9\mu^2 - 4)\, M_o^8 + (4\mu^4 + 4\mu^2 + 11)\, M_o^6 + \\ +(-6\mu^4 - 10\mu^2 + 10)\, M_o^4 + (4\mu^4 - 4\mu^2 - 1)\, M_o^2 - \mu^4 + \mu^2$$

which lead to :

(57)

$$\frac{1}{\rho_1} + \frac{1}{\rho_2} = \omega + \mathcal{O}\left(\frac{1}{\mathcal{M}^2}\right)$$

$$\frac{1}{\rho_1 \rho_2} = \pi + \Theta\left(\frac{1}{\mathcal{M}^2}\right)$$

CONCLUSION :

The presentation of an application of the ε-algorithm to a well-determined physical problem demonstrates the difficulty of the choice between such and such approximation.

The advantage of the acceleration processes results from the fact that they permit to place oneself immediately closest to the accurate solution and the initial value of numerical calculations.

In the preceding formulae (distance of vertices, curvature at the vertex) are applied to the case of a spheric or paraboloidal body with a leading angle, equal to 0, comparaison with the numerical results obtained from computers through longer methods demonstrates that the convergence of approximations seems to be correct and that even with reference to curvature without being excellent the first order approximation gives reasonable results.

The method used here is a direct one, the shock being obtained from the given body, we have limited ourselves to first and second order approximations, the parameters of the problem remaining indeterminate.

It must be noted that PADE'S methods applied up to the twelfth order by computer lead to remarkable results for the inverse problem. They are comparable to the results obtained here in the considered cases, i.e for a perfect diatomic gas, an infinite upstream number of Mach and on axisymmetrical flow, the chosen shock surface being a paraboloid.

K.I. BABENKO, G.P. VOSKRESENSKII, A.N. LYUBIMOV and
V.V. RUSINOV - Spatial Flow of an Ideal Cas around smooth Bodies - Moscow 1964

G.A. BAKER The theory and application of the PADE method - Adv. Theo. Phys. 1 1965

H. CABANNES Tables pour la détermination des ondes de choc détachées La recherche aéronautique 36 - Paris 1953

J.P. GUIRAUD Possibilités et limites actuelles de la théorie des écoulements hypersoniques - O N E R A Publ. 99 - Paris 1961.

M.M. HAFEZ and H.K. CHENG
On acceleration of convergence and shock-fitting in transonic flow computation - Univ. of Southern - California - 1973.

H. PADE Sur la représentation approchée d'une fonction par des fractions rationnelles. Ann. Sci. Ecole. Norm. Sup. 9. Paris 1892

L.W. SCHWARTZ Hypersonic flows generated by parabolic and paraboloidal shock waves. N.A.S.A. 1973

D. SHANKS Non-linear transformations of divergent and slowly convergent sequences - J. Math. and Phys. 34 - 1955

M. VAN DYKE and H. GORDON

Supersonic flow past a family of blunt axisymmetric bodies - N.A.S.A. Tech. Report 1959.

A. VAN TUYL — Use of rational approximations in the calculation of flows past blunt bodies - A.I.A.A. 5 - 5 1967.

P. WYNN — The rational approximation of functions which are formally defined by a power series expansion - Math. of Comp. 14 -1960

CYCLIC ITERATIVE METHOD APPLIED TO TRANSONIC FLOW ANALYSES*

H. K. Cheng, University of Southern California
Los Angeles, California

and

M. M. Hafez, Flow Research, Inc.
Los Angeles, California

SUMMARY

This paper reviews recent works on acceleration techniques for iterative solutions of elliptic and mixed-type problems, using algorithms related to Padé's fractions. The study focuses on the question of how to speed up convergence of relaxation methods currently available for transonic and related flow computations, with minimal alterations in computer programming and storage requirements.

The theoretical basis of the work is similar to the power method, but allowance is made that moduli of some of the eigen-values can be very close to one another and to unity. The study contributes to a clarification of the error analyses for the sequence transformations of Aitken, Shanks, and Wilkinson, and to developing a cyclic iterative procedure applying the transformations to accelerating large linear and nonlinear systems. Use of the first and second order transforms similar to Shanks' (corresponding to the second and third rows in the upper half of Padé's Table) is shown to be effective, but their subtle differences from the latter prove to be crucial.

Examples illustrating the accelerating technique include transonic flow as well as model Dirichlet problems. Reduction by a factor of three to five in computing time is possible, depending on the accuracy requirement and the order of the transformation. The possibility for reducing the computer storage requirement via Wynn's recursive identities is examined for a linear system in Appendix A.

*This research was supported by the Office of Naval Research under Contract Number N00014-67-A-0269-0021.

I. INTRODUCTION

Many current computation methods in fluid dynamics make use of relaxation procedures, wherein solutions are obtained after a sufficiently large number of iterations. One recent advance in this respect is, perhaps, the calculation of plane transonic flow by Murman & Cole [1] and the subsequent extensions by many workers. (For example, see Refs. 2-4.) These methods, using type-dependent schemes in the discretization and following line-relaxing procedures, succeed in capturing shock waves in supercritical flows. The computer storage and the number of operations of the programs are low enough to make the computation possible even for a modest institution. However, the computer time of 400 - 1000 iterations required for the more complicated problems may still demand 1/2 to 2 hours on an IBM 360 (or 370), and 10 - 40 minutes on a CDC 6600. Use of acceleration technique with a savings in computer time by a factor of 3 or 4 is certainly worthwhile, especially if one has a great number of problems to solve.

The convergence rate of these relaxation procedures will depend on the largest eigen-values of the iterative matrix, λ_1, referred to subsequently as the spectral radius. The error (norm) at the kth iteration is, in most cases, gauged by $|\lambda_1|^k$. The need of acceleration follows from a rather well-known fact that the spectral radius tends to unity, as the mesh size vanishes. (See for example, Refs. 5-8.)

The central question in the following is, therefore, how to speed-up convergence of relaxation methods currently available for the transonic flow and similar computations with minimal alteration in computer programming and storage requirements.

Our acceleration technique is basically a cyclic iterative procedure; transformations related, but, not identical, to the nonlinear transforms of Shanks and Aitken, and to the Padé fraction are applied at the conclusion of each iterative cycle.

Although the genesis of our study may be traced back to the Padé fraction, its rational basis is derived from the power method of Fadeev & Fadeeva [8], but, special allowance is made in the error analysis that the magnitudes of some of the eigen-values can be very close to unity, and to one another, and may also repeat themselves.

An algorithm using Padé's fraction has been employed recently by Martin & Lomax to accelerate their relaxation method for transonic flow. [9,10] Their basic iterative procedure takes advantage of a fast, elliptic solver; but, it still makes use of the type-dependent schemes similar to other line-relaxation programs (this is, itself, quite novel). In their acceleration procedure, however, a three-term expansion in an artificial parameter ε is used; solutions to a 3rd-order perturbation problem are then used to generate the Padé fraction (at $\varepsilon = 1$). A savings in computer time by a factor of nearly two has been reported for certain (but not all) cases studied.

Much of the material used below is taken from our work on "Convergence Acceleration and Shock Fitting for Transonic Aerodynamics Computations", AIAA paper 75-51. An updated version of this work has been distributed as an University of Southern California Report.[11] This talk will concern primarily the acceleration technique, and we shall take this opportunity to examine more closely the underlying ideas, and their subtle differences from those of Shanks[12] & Padé[13]. What we hope to convey in the following is that only the most elementary of the transforms and their equalities have been used; they, nonetheless, have been very helpful.

II. REMARKS ON PADÉ FRACTION AND SHANKS SEQUENCE TRANSFORMATIONS

The use of transformation to improve the convergence characteristics of sequences is a recurrent theme of this proceeding. One class of transforms, which bears a close relationship to the key equations in our method, is that of Shanks and the corresponding Padé fractions. [12,13,&14] The simplest among these is the "e_1" transform (singled-out in block on the left side of the diagram inserted below), which predicts the convergence limit ϕ from three successive iterates ϕ_{k-1}, ϕ_k, and ϕ_{k+1}. This formula has a long history and is referred to in some quarter as the δ^2 - process of Aitken.[14] Considering ϕ_k as the k-term partial sum of a series for an analytic function, Shanks identifies one of his transformed sequences e_n of ϕ_k (to which the "e_1" belongs) with that of the rth row of the Padé table above the diagonal.‡

SHANK'S HIERARCHY

$$\begin{array}{cccccccc} e_1 & e_2 & e_3 & \cdot & \cdot & e_n & \cdot & \cdot \\ e_1^2 & e_2^2 & e_3^2 & \cdot & \cdot & e_n^2 & \cdot & \cdot \\ & & & & & & & \\ \cdot & \cdot & \cdot & \cdot & & e_n^p & \cdot & \cdot \end{array}$$

$$\boxed{e_1(\phi_k) = \frac{\phi_{k-1}\phi_{k+1} - \phi_k^2}{\phi_{k-1} - 2\phi_k + \phi_{k+1}}}$$

$$e_n(\phi_k) = \sum_{i=1}^{n} c_i \phi_{i+k} \Big/ \sum_{i=1}^{n} c_i$$

PADE FUNCTION

$$\begin{array}{cccccccccc} \phi_0 & \phi_1 & \phi_2 & - & - & - & \phi_{m-1} & \phi_m & \phi_{m+1} & - \\ r_{01} & r_{11} & r_{21} & - & - & - & - & r_{m1} & - & - \\ r_{02} & r_{21} & r_{22} & - & - & - & - & r_{m2} & - & - \\ & & & & r_{nn} & - & - & r_{mn} & - & - \end{array}$$

$$\boxed{e_n(\phi_k) = r_{kn}}$$

$$\phi(t) = b_0 + b_1 t + b_2 t^2 + \cdots$$

$$r_{mn} = P_m(t;n)/Q_n(t;m)$$

‡In Shanks' original paper the e_n is written as e_k, with k denoting the order of the transform.

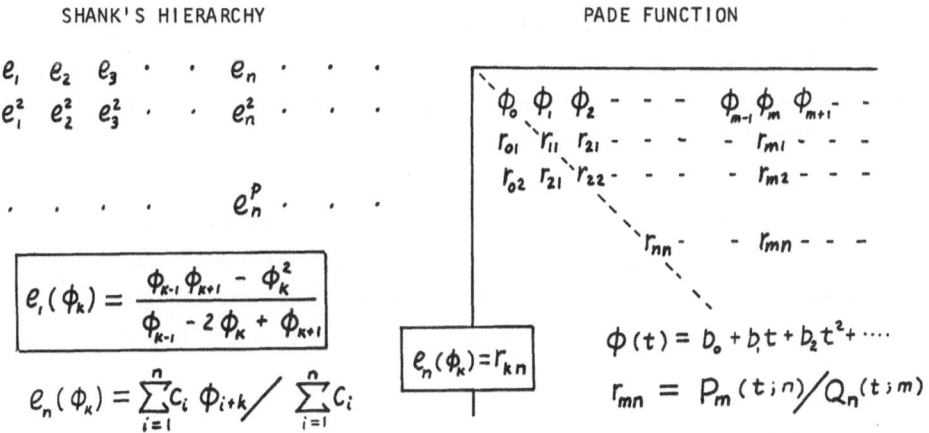

Much work has been done in uncovering the many important and interesting properties of, and identities among, the elements in the Padé Table, hence, the e_n-sequences. (See for example, Refs. 15-17.) But much has yet to be learned about the error estimates in general applications. A rather superficial remark one could give in this regard is this: If the ϕ_k's are the iterative solutions to a nonlinear scalar equation

$$\phi_{k+1} = g(\phi_k) ,$$

then the e_1 transform predicts the limit ϕ with an error comparable to the square of the error in the original sequence. The transform in this case is simply a derivative-free variant of Newton's method. But, this superlinear accuracy does not hold for a system of equations involving more than one unknown; the accuracy of the transforms in this case must be established on a different basis. There is a second observation related to the accuracy of the e_n transform, which is also quite well known after Shanks original work[12], namely, the transform e_n of ϕ_k represents the exact limit ϕ, if the sequence ϕ_k has precisely the transient behavior for successive k in the exponential involving k^{th} powers of q_i's

$$\phi_k = \phi + \sum_{i=1}^{n} \alpha_i q_i^k .$$

It is apparent that convergence will require the magnitude of each q_i to be less than one; it also follows that the prediction would be exactly correct, if the sequence ϕ_k happens to be the partial sum of n geometric series.

The stipulated exponential transient cannot be one of general validity, because there is no a priori reason that the iterates of a general scalar equation should not approach its limit algebraically! However, for iterative solutions to a system of algebraic equations of interest, a similar exponential transient does apply to each component of the solution near the convergence limit. This basis is provided by the power method to be discussed below in Section 3.

In passing, we may observe that, owing to the storage limitation, application of the e_n or equivalent transform beyond e_1 and e_2 may not be easily accommodated in a computer program. Therefore, our acceleration scheme has been limited to these corresponding to elements far removed from the diagonal in the Padé Table. This is in contrast to most applications of Páde fraction in fluid mechanics today.[17-20]

III. THE POWER METHOD

In our application involving a large algebraic system, the unknown is the velocity potential ϕ, and its kth iterate is ϕ_k. They may be considered as "vectors" with components as many as the number of total grid points used in the relaxation method, say N. The new iterate at the (k+1) iteration, is a function of the "vector" ϕ_k

$$\phi_{k+1} = g(\phi_k)$$

determined by the difference schemes and the iterative procedure used. We regard ϕ_k as a perturbed solution from the convergence limit

$$\phi_k = \phi + \epsilon_k , \qquad \epsilon_k \to 0.$$

Near the limit, the nonlinear iterative equation yields a linear, recursive relation for the error vector

$$\epsilon_{k+1} = Q\, \epsilon_k$$

where Q is the Jacobian matrix of the function g of ϕ , independent of k and ϕ_k. If the eigen-values of this matrix λ_i's are distinct, we may represent the initial error vector ϵ_0 by a linear combination of the eigen-vectors

$$\epsilon_0 = \sum_{i=1}^{N} \alpha_i v_i .$$

This leads to a form for the error vector

$$\epsilon_k = \sum_{i=1}^{N} \alpha_i v_i \lambda_i^k .$$

It shows that the error vector decays (or amplifies) exponentially in k. For convergence, the magnitudes of the eigen-values must be less than one, just like the requirement on the q_i's in Shank's exponential transient. This is the main base for the power method of Fadeev & Fadeeva, as well as the two of our transformations to be discussed below.

The linear recursive equation for the error vector has an exact analog in the discretized version of the time-dependent system

$$C\dot{\varphi} = A\varphi , \qquad \text{with} \qquad Q = \exp.(\Delta t\, C^{-1}A) .$$

From this, we may see the prospect for accelerating a pseudo-unsteady fluid mechanics problem.

Let us adopt Fadeev & Fadeeva's result and order the eigen-values according to their absolute magnitudes

$$|\lambda_1| > |\lambda_2| > |\lambda_3| > \cdots > |\lambda_i| > |\lambda_{i+1}| > \cdots > |\lambda_N| .$$

After long enough iterations, i.e., large enough k, one may omit all but one term associated with the first eigen function v_1. This, after eliminating $\alpha_1 v_1$, leads to our first-order transform, which predicts the limit ϕ from two successive iterates with an unknown eigen-value λ_1 . If one chooses to eliminate λ_1 by three successive iteraties, he will recover the e_1-transform corresponding to the first row of the Pade Table

$$\phi = \phi_K + \frac{\phi_{K+1} - \phi_K}{1 - \lambda_1} .$$

However, more useful estimates of λ_1 are obtained by averaging over all components through proper sum and inner products illustrated as, with $\delta_K = \epsilon_K - \epsilon_{K-1}$,

$$\lambda_1 = \sum^{N} |\delta_{K+1}| \Big/ \sum^{N} |\delta_K| \text{ , or } \overline{\overline{\lambda}}_1 = \delta_K^T \delta_{K+1} \Big/ \delta_K^T \delta_K .$$

A justification for $\overline{\overline{\lambda}}_1$, is given in Appendix B.

To generate formula corresponding to the e_2 and higher-order transforms, we simply include more and more higher-order eigen functions into the "transient" representation, and obtain

where $p_n = 1$ and pj's are constants for the entire field. One of our modest contributions, delineated in Ref. 11, is to analyse the remainder and confirm the classical results for repeated and closely spaced eigen values.

IV. APPLICATION TO A CYCLIC ITERATIVE PROCEDURE

In the application, the transforms are used as a part of an iterative algorithm: the procedure consists of several cycles, each makes k' iterations (say 10 - 30). The transform is applied at the end of each cycle to yield an estimate of the limit, to be used as initial data for the next cycle. The sketch in Fig. 1 illustrates the method when the first-order transform is used, which needs data from three stages of iterations. Note that these three values can be taken from values at k-m, k, and k+m, for some integral m. Additional storage for whole sets of field data is required, and it varies from 1 to 4 sets, depending on the order of the transform and ways the eigen-value estimates are handled.

In passing, we note that if the δ^2-process is strictly applied for each component, i.e. at each grid point, not only more storage is required but the redundant eigen-value estimates implicit in such process would lead to inconsistency and delay the approach to the limit. We find the δ^2-process coverges much more slowly in most cases.

V. EXAMPLE: A DIRECHLET PROBLEM

In a study described in our report[11], we have tested this cyclic-transform technique on line-relaxation methods applied to a model Dirichlet problem, using various relaxation parameters and sweep directions. These numerical experiments show that a reduction in iteration number by a factor of three to five is generally possible. In the example of Fig. 2, a line Gauss-Seidel procedure is applied to the system based on a 9-point central difference scheme, for which neither the optimum relaxation parameter, nor the spectral radius, is theoretically known to the best of our knowledge. A typical convergence history of the unaccelerated results, using a 1/30th mesh, is shown as a solid curve. The abscissa of the graph is the iteration number k. The accelerated result based on a 2nd-order transform is shown in short dash with circles, which approaches the limit within 1% in 30 iterations, as compared to three to four hundred for the unaccelerated one.

VI. EXAMPLES: TRANSONIC THIN AIRFOIL PROBLEMS

We shall study below the results of application in transonic small-disturbance theory governed by the von Kármán equation.[21] The discussion is confined to the flow over a symmetric circular arc airfoil, which has an embedded supersonic region. The basic program to be accelerated is one similar to that of Murman and Cole,[1] using an x-mesh of 2½% chord, and a y-mesh near the wing 2% chord. For the result shown in Figure 3a, the relaxation parameter is taken to be 1.4 in the subsonic region and 0.9 in the supersonic region. This slide gives the convergence history for the velocity perturbation near the mid chord. The unaccelerated result shown in solid curve takes 140 iterations to approach the limit within 1%; cyclic acceleration using the first-order transform presented in thin solid curve takes 60 iterations for the same accuracy. For the results using 2nd-order transform, only data at the end of each cycle are shown in circles; this takes only 40 iterations to reach the limit within 1%.

The results in Figure 4b differ from the preceeding one in that, here, a uniform relaxation parameter, $\omega = 0.95$, is used in the supersonic and subsonic regions. The convergence rate for the unaccelerated program in solid curve is low, as expected, taking 400 iterations or more to reach the limit within 1%. This is to be compared with the 65 and 30 iterations for the two accelerated solutions.

We have also studied the acceleration of transonic solutions involving circulation, i.e. airfoil at incidence. The basic line-relaxation program is the same as before, except for a doubling in the number of grid points to account for the asymmetry and the use of a somewhat different pair of relaxation parameters. One sees from Fig. 5 that the use of the first-order transform in solid curve achieves a convergence within 1% at 150 iterations for the circulation, whereas the unaccelerated one may take more than 400.

We would like to emphasize that the above examples involve shock waves which are "captured", so to speak, by the numerical procedure - thanks to the "numerical viscosity" inherent in the computer program. Because of this, the flow detail near the shock is lost. A shock-fitting method, which modifies the computer program to fit the shock as a surface of discontinuity, has been developed.[11] The natural question to be asked is whether acceleration and shock-fitting techniques can work together. The answer is an affirmative one. In Figure 5, we present the shock and sonic boundaries from our iterative, shock-fitting solution, computed for a slightly supersonic Mach number. The unaccelerated result obtained after 240 iterations compares well with that obtained by Magnus & Yoshihara, who used a shock capturing method based on an unsteady approach. With acceleration based on a 2nd-order transform, the very same shock-fitting solution is recovered in 64 iterations.

VII. CONCLUDING REMARKS

In summary, our study with the transonic flow and other examples show that the cyclic acceleration techniques based on sequence transforms may effectively increase the convergence rate and the efficiency of the relaxation methods, with minimal programming and storage changes. A reduction by a factor of three to five in computer time is possible, with and without shock-fitting. In fact, where accurate description for the shock is important, the time saved by acceleration with shock fitting can be 6 to 36 fold. One observes that the above demonstration involves only the use of some of the most rudimentary forms of sequence transforms. With an increase in data storage capacity (or facility), it should be possible to employ the more sophisticated higher order transforms and their recurrence relations which are discussed in other parts of this Proceeding. In the meantime, possibilities for reducing the data storage requirement for the higher-order transforms do exist. This is supported by a study described in Appendix A below for an iterative procedure applied to a linear system, making use of Wynn's recursive relations for the ε-algorithm.

APPENDIX A. IMPLEMENTATION OF WYNN'S ε-ALGORITHM FOR APPLICATIONS TO ITERATIVE MATRIX EQUATIONS

In Ref. 22, Wynn uses the ε-algorithm as an acceleration technique for iterative vector and matrix problems.[+] The effective use of the transforms in the cyclic iterative method discussed in the text, as well as the corresponding elements in the ε-algorithm, are limited, in practice, by the increased storage requirement for the higher-order transforms. However, the possibility for using higher-order transforms without the increasing storage remains, and is confirmed below for a linear system. This is accomplished through application of Wynn's

[+]The symbol "ε" employed in this Appendix, is not to be confused with the error vector "ϵ_k" used in the text.

rhombus rule, and other identities for a linear iterative equation system. The result provides an alteration from Wynn's original procedure [22] with a substantial savings in data storage.

Wynn's Recursive Relation Applied to Vectors and Matrices

The power of the ε-algorithm lies in the fact, established through many examples, that if the sequence $\phi_0, \phi_1, \phi_2, \ldots, \phi_k, \ldots\ldots$, i.e.,

$$\{\phi_k\} = \varepsilon_0^{(k)}$$

is slowly convergent, then the numerical convergence of the sequence $\varepsilon_0^{(0)}, \varepsilon_2^{(0)}, \varepsilon_4^{(0)}, \ldots, \varepsilon_{2s}^{(0)}, \ldots$, i.e.,

$$\{\varepsilon_{2s}^{(0)}\}$$

to the limit (or antilimit), with which sequence $\{\phi_k\}$ is associated, is far more rapid. In the scalar case, the quantities $\varepsilon_s^{(k)}$ satisfy the rhombus rule [15]

$$\varepsilon_{s+1}^{(k)} = \varepsilon_{s-1}^{(k+1)} + \frac{1}{\varepsilon_s^{(k+1)} - \varepsilon_s^{(k)}}, \tag{A.1}$$

which is closely related to the Shanks' transform. As is well known, the elements generated in this manner in the ε-algorithm may be identified with those on the upper half of the Pade Table (cf. Sec. II in text), hence, those in Shanks' e_n transform,

$$\left.\begin{aligned} \varepsilon_{2n}^{(k)} &= r_{k+n,n} \\ \varepsilon_{2n+1}^{(k)} &= r_{k+n+1,n} \end{aligned}\right\} \tag{A.2}$$

with $\varepsilon_0^{(k)} = \phi_K$, and setting $\varepsilon_{-1}^{(k)} = 0$.

In cases in which ϕ_k is a vector or a matrix, the algorithm is still meaningful, provided the inverse of the entity is consistently defined. Wynn has considered the following alternative definitions.

(i) Primitive Inverse: In this case, each component is considered independently; it amounts to a simultaneous application of the scalar ε-algorithm to components of the array.

(ii) The Samelson Inverse of a Vector: In this case, the inverse of the vector $X = (x_1, x_2, \cdots, x_N)$ is taken (after K. Samelson) to be

$$X^{-1} \equiv \left(\sum_{j\ 1}^{N} x_j\, x_j\right)^{-1} (\bar{x}_1, \bar{x}_2, \cdots \bar{x}_N), \tag{A.3}$$

where $\bar{x}_j$ is the complex conjugate of x_j.

(iii) The Normally Defined Inverse of a Square Matrix: This was not recommended for large systems.

Wynn discusses in Ref. 22 applications of the algorithm to numerical analyses, including boundary-value problems, initial-value problems, Fredholm and Voltera integral equations, and differential equations.

Wynn's Procedure for Accelerating Relaxation Solutions

Of particular interest are Wynn's application to the acceleration of the Jacobi and Gauss-Seidel relaxation methods for iterative solution of large systems of linear algebraic equations.[22]

For subsequent discussion, it is convenient to arrange the array of $\varepsilon_s^{(k)}$ into the familiar pattern suggested by the rhombus rule, illustrated at the middle of the page, where the original sequence $\{\phi_k\}$ is given on the first non-zero column near the left. With the identification given by Eq. (A.2), elements on each column corresponds to those belonging to Shanks' e_n- transform of the same order (with the order increasing towards the right). The diagonal elements in the Padé Table $r_{n,n}$ are identified with elements on the "roof top" with even subscript, i.e., with $\varepsilon_{2s}^{(0)}$.

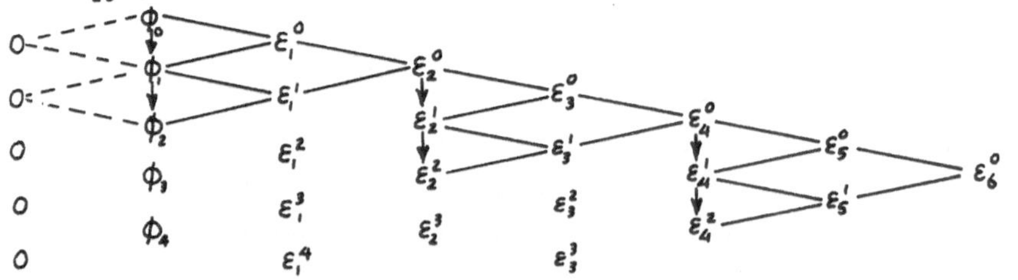

In Wynn's applications, a sequence of vectors or matrices is obtained from an iterative procedure for a linear system, say,

$$\phi_{k+1} = Q\phi_k + b \ , \tag{A.4a}$$

and stored as $\varepsilon_0^{(k)}$ before the acceleration procedure is applied. For example, if we have three iterates ϕ_0 , ϕ_1 , and ϕ_2 , a better estimate will then be determined as $\varepsilon_2^{(0)}$ according to the rhombus rule (which in this case is identifiable with Padé's r_{11} or Shanks' $e_1\{\phi_1\}$). If more resolution is needed, i.e., if one wishes to obtain $\varepsilon_n^{(0)}$, with $n \geqslant 4$, more iterates (with $k \geqslant 4$) have to be generated from Eq. (A.4) and stored.

Underlying this procedure is the assumption that at the end point (towards the right) of the application of the rhombus rule, one shall arrive at (or near) the limit ϕ satisfying the equation

$$\phi = Q\phi + b . \tag{A.4b}$$

This assumption can indeed be justified. In fact, inasmuch as the number of components of ϕ_k , say N, is finite, the exact solution ϕ can be predicted from ϕ_0 and 2N (and only 2N) successive iterates, i.e., ϕ_0 , ϕ_1 , $\phi_2, \dots, \phi_k, \dots \phi_N$,

$\phi_{N+1}, \cdots, \phi_{2N-1}, \phi_{2N}$, using rhombus rule. This follows from Eq. (A.4a), for which a corollary of the Cayley-Hamilton theorem (cf. Eq. (3.9) on p. 5, of Ref. 11) gives †

$$\phi = \phi_k + \sum_{j=0}^{N} p_j (\phi_{k+j} - \phi_k) \Big/ \sum_{j=0}^{N} p_j \,, \tag{A.5a}$$

where pj's are the coefficients in the characteristic polynormal of the iterative matrix Q

$$\prod_{j=1}^{N} (\lambda - \lambda_j) = p_0 + p_1\lambda + p_2\lambda^2 + \cdots + p_N\lambda^N. \tag{A.5b}$$

Now, the right hand member of Eq. (A.5a) is precisely $e_N(\phi_k)$, identifiable with $\varepsilon_{2N}^{(k-N)}$. Hence, the N-component vector ϕ is precisely recovered from any 2N+1 consecutive iterates of Eq. (A.4a) upon reaching the end of application of the rhombus rule (at the 2Nth column of the $\varepsilon^{(k)}$ array, for any k). Note that the validity of Eqs. (A.5a) and (A.5b), hence, the conclusion, requires the eigen-values, λ_j's, to be neither distinct nor completely real. On the same basis, it is also possible to recover the entire set of the eigen-values from the ratio $\varepsilon_s^{(k+1)} / \varepsilon_s^{(k)}$ in the limit $k \to \infty$.

The above shows that the application of the ε-algorithm amounts to providing a 2N finite steps process for solving a set of N linear equations. This would require, however, the storage of $2N^2$ pieces of data, which may not be desirable for a large system.

<u>Generating $\varepsilon_{2s}^{(k)}$ By Iterations</u>

Our procedure for applying the (higher-order) ε-algorithm without the penalty of an increased computer storage relies on a theorem of the ε-array. Namely, if the successive (vector) ε-elements on the first (non-zero) column are generated by a linear matrix iterative law (equation), say,

$$\phi_{k+1} = Q\phi_k + b \,, \tag{A.6}$$

successive vector elements on any other even column obey, and can be generated from, the <u>same</u> iterative law, i.e.,

$$\varepsilon_{2s}^{(k+1)} = Q\, \varepsilon_{2s}^{(k)} + b \,. \tag{A.7}$$

This special aspect concerning linear iterative systems was not considered in Wynn's work.[22] The following proof through induction makes use of a recursive relation of Wynn [15] corresponding to the "missing identity of Frobenius".

† In Eq. (3.8b) of Ref. 11, "i = 1" should be written as "i = n+1".

Let E, W, N, and S denote the four elements around an element C in the ε-array, in the order of East, West, North, and South. Wynn's recursive identity for elements of the even column is[15]

$$\frac{1}{N-C} + \frac{1}{S-C} = \frac{1}{E-C} + \frac{1}{W-C} , \tag{A.8}$$

or,

$$E = \frac{1}{\frac{1}{N-C} + \frac{1}{S-C} - \frac{1}{W-C}} + C . \tag{A.9}$$

To render (A.8) or (A.9) applicable at the second even column (from the left), one may introduce an additional column with $\varepsilon_{-2}^{(k)} \to \infty$, which is consistent with the auxiliary column $\varepsilon_{-1}^{(k)} = 0$. Let the superscript $*$ refer to a successive iterate, e.g. $\phi_k^* = \phi_{k+1}$. It will be established first that if the iterative law Eq. (A.7) holds in two neighboring even columns, it will also hold in the next even column. In particular, we want to show that, if Eq. (A.7) is applicable to W, N, S, and C, it will also hold at E which is related to the others through rhombus rule, or better through Eq. (8) or (9). Now, multiply both sides of Eq. (A.9) by Q, and add b; we have

$$QE + b = Q\left(\frac{1}{\frac{1}{N-C} + \frac{1}{S-C} - \frac{1}{W-C}}\right) + QC + b. \tag{A.10}$$

But the first term on the right is

$$Q\left((N-C)^{-1} + (S-C)^{-1} - (W-C)^{-1}\right)^{-1}$$

$$= \left[(Q(N-C))^{-1} + (Q(S-C))^{-1} - (Q(W-C))^{-1}\right]^{-1} \tag{A.11}$$

If we assume that Eq. (A.7) hold at N, S. W, and C, then the R.H.S. of Eq. (A.10) becomes

$$\frac{1}{\frac{1}{N^*-C^*} + \frac{1}{S^*-C^*} - \frac{1}{W^*-C^*}} + C^* ,$$

which is, according to the recursive relation Eq. (A.9), E^*. Hence, Eq. (A.10) yields

$$E^* = QE + b , \tag{A.12}$$

confirming that Eq. (A.7) holds along the next even column (to the right of the two even columns containing N, S, C, and W). It remains to show that the same iterative law applies along the first two even columns involving elements

$$\varepsilon_{-2}^{(k)} \to \infty , \qquad \varepsilon_0^{(k)} = \phi_k .$$

This is true under Eq. (A.6). Hence,

$$\varepsilon_{2s}^{(k+1)} = Q\,\varepsilon_{2s}^{(k)} + b\,, \tag{A.13}$$

provided it holds for $s = 0$.

Departure From Wynn's Original Procedure: Significance

The significance of Eq. (A.13) with the accompanying provision lies in the fact that a part of the elements on even columns in the ε-array may now be generated alternatively by iteration. In other words, use of iterations along successive even columns may be exchanged with the storage required for $\varepsilon_0^{(k)} = \phi_K$, which is large for the higher order transforms. We note in passing that the matrix operations in Eq. (A.11), hence, the theorem, holds also in cases where the $(N-C)^{-1}$, etc., are defined by the "primitive inverse" and other alternatives mentioned earlier.

With the theorem Eq. (A.13), any element in the ε-array can be recovered from three initial iteraties $\varepsilon_0^{(0)} = \phi_0$, $\varepsilon_0^{(1)} = \phi_1$, and $\varepsilon_0^{(2)} = \phi_2$, applying the same iterative law to generate elements in the intermediate even columns. This procedure is illustrated in the diagram above Eq. (A.4a) for a case in which the end point is $\varepsilon_6^{(0)}$; the downward arrows indicate generation of elements by iterations, the net-work otherwise signifies application of the rhombus rule, Eq. (A.1), in the usual manner. The most crucial feature of such a procedure is, perhaps, the fact that, in the process of generating intermediate elements in the even and odd columns, the data storage never exceeds the requirement for storing three pieces of $\varepsilon_s^{(k)}$, for any s and any k. We note that, a similar procedure based on Eq. (A.8) alone may also be used to recover elements in the even column; this however requires storage for four ε's instead of three as in the one described above.

This modified ε-algorithm may be used in the cyclic iterative procedure for relaxation methods described in the text, in which the algorithm will be applied to predict the limit from successive iterates, but will no longer be handicapped by the excessive storage requirement. One potential application is to use the procedure continuously (in one long cycle), corresponding to an unbroken ZIG-ZAG path along the "roof top" of the ε-array. This reprents a new relaxation procedure, of which the problems of stability and rounding errors deserve attention in future study.

APPENDIX B. ALTERNATIVE CRITERION FOR DETERMINING THE RELAXATION PARAMETERS

Consider a nonlinear relaxation equation

$$\phi_{K+1} = f(\phi_K). \qquad \text{(B.1)}$$

Let us assume

$$\phi = \phi_K + \omega_1(\phi_K - \phi_{K-1}) + \\ + \omega_2(\phi_{K-1} - \phi_{K-2}) + \cdots ,$$

and let $J(\phi)$ be a functional associated with Eq. (B.1). Then the stationary condition

$$\delta J(\phi) = 0$$

should provide a system of equations for the unknown parameters ω_1, ω_2, etc., Eq.(B.1) may represent, for example, a minimum error principle. For a linear system, whose iterative matrix has a dominant eigen-value, the ω_1 obtained from least squares criterion is identical with the inner product form of the λ_1, i.e., $\bar{\bar{\lambda}}_1$ in our work.[11]

ACKNOWLEDGEMENT

This paper is based on a study at the University of Southern California supported by the Office of Naval Research, Fluid Dynamics Program, under contract number N00014-75-C-0520.

REFERENCES

1) Murman, E. M. and Cole, J. D., "Calculation of Plane Steady Transonic Flow," AIAA Jour., Vol. 9, no. 1, 1971, pp. 114-121.

2) Krupp, J. A. and Murman, E. M., "Computation of Transonic Flows Past Lifting Airfoils and Slender Bodies," AIAA Jour., Vol. 10, No. 7, 1972, pp. 880-886.

3) Garabedian, P. R. and Korn, D. G., "Numerical Design of Transonic Airfoils", in Numerical Solution of Partial Differential Equations - II, Academic Press, 1971.

4) Jameson, A., "Numerical Calculation of the Three Dimensional Transonic Flow over a Yawed Wing", Proceedings AIAA Computational Fluid Dynamics Conference, pp. 18-26.

5) Young, D., Iterative Solutions for Large System of Linear Equations, Academic Press, New York, 1971

6) Varga, R. S., Iterative Matrix Analysis, Prentice-Hall, Englewood Cliffs, New Jersey, 1962.

7) Wilkinson, J. H., The Algebraic Eigen-value Problem, Clarendon Press, Oxford, 1965.

8) Fadeev, D. K. and Fadeeva, V. N., Computational Methods of Linear Algebra (translated by R. C. Williams), W. H. Freeman & Co., San Francisco, 1963.

9) Martin, E. D. and Lomax, H., "Rapid Finite Difference Computation of Subsonic and Transonic Aerodynamic Flows", AIAA paper No. 74-11, 1974

10) Martin, E. D., "Progress in Application of Direct Elliptic Solver to Transonic Flow Computations", to appear in Aerodynamic Analyses Requiring Advanced Computers, NASA SP-347, 1975.

11) Hafez, M. M. and Cheng, H. K., "Convergence Acceleration and Shock Fitting for Transonic Aerodynamics Computations", Univ. So. Calif., School of Eng'r., Rept. USCAE 132, April 1975.

12) Shanks, D., "Nonlinear Transformations of Divergent and Slowly Convergent Sequences", Studies of Applied Math., (J. Math. Phys.), No. 34, pp. 1-42, 1955.

13) Padé, H., "Sur la representation approchee d'une fonction par des rationelles", Ann. Ecole Nor (3), Supplement, 1892, pp. 1-93.

14) Aitken, A. C., "Studies in Practical Mathematics II, Proc. Royal Soc. Edinburgh, Vol. 57, 1937, pp. 269-304.

15) Wynn, P., "Upon System of Recursions Which Obtain Among the Quotients of the Padé Table", Numer. Math., Vol. 8, 1966, pp. 246-269.

16) Gragg, W. B., "The Padé Table and Its Relation to Certain Algorithms of Numerical Analysis", SIAM Review, Vol. 14, No. 1, 1972.

17) Baker, G. A., Gammel, J. L., and Willis, J. G., "An Investigation of the Applicability of the Padé Approximation Method", Jour. Math. Anal., Vol. 2, 1961, pp. 405-418.

18) Van Dyke, M. D., "Analysis and Improvement of Perturbation Series", Quart. Jour. Mech. Appl. Math., Nov. 1974.

19) Van Tuyl, A. H., "Calculation of Nozzle Using Padé Fractions", AIAA Jour., Vol. 11, No. 4, 1973, pp. 537-541.

20) Cabannes, H. and Bausset, M., "Application of the Method of Padé to the Determination of Shock Waves", in Problems of Hydrodynamics and Continuum Mechanics, in Honor of L. I. Sedov, English ed. published by SIAM, 1968, pp. 95-114.

21) von Kármán, T., "The Similarity Law of Transonic Flow", Jour. Math. and Physics, Vol. 26, 1947, p. 3.

22) Wynn, P., "Acceleration Techniques for Iterated Vectors and Matrix Problems", Math. Comput., Vol. 16, 1962, pp. 301-322.

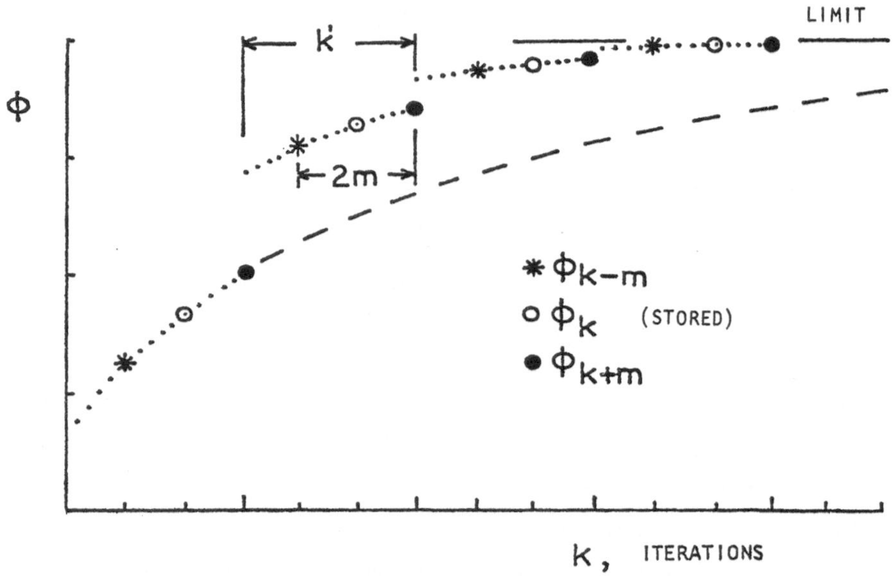

Figure 1. Cyclic acceleration technique applied to an iterative solution, illustrated for the first-order transform (the cycle is repeated every k' iteration).

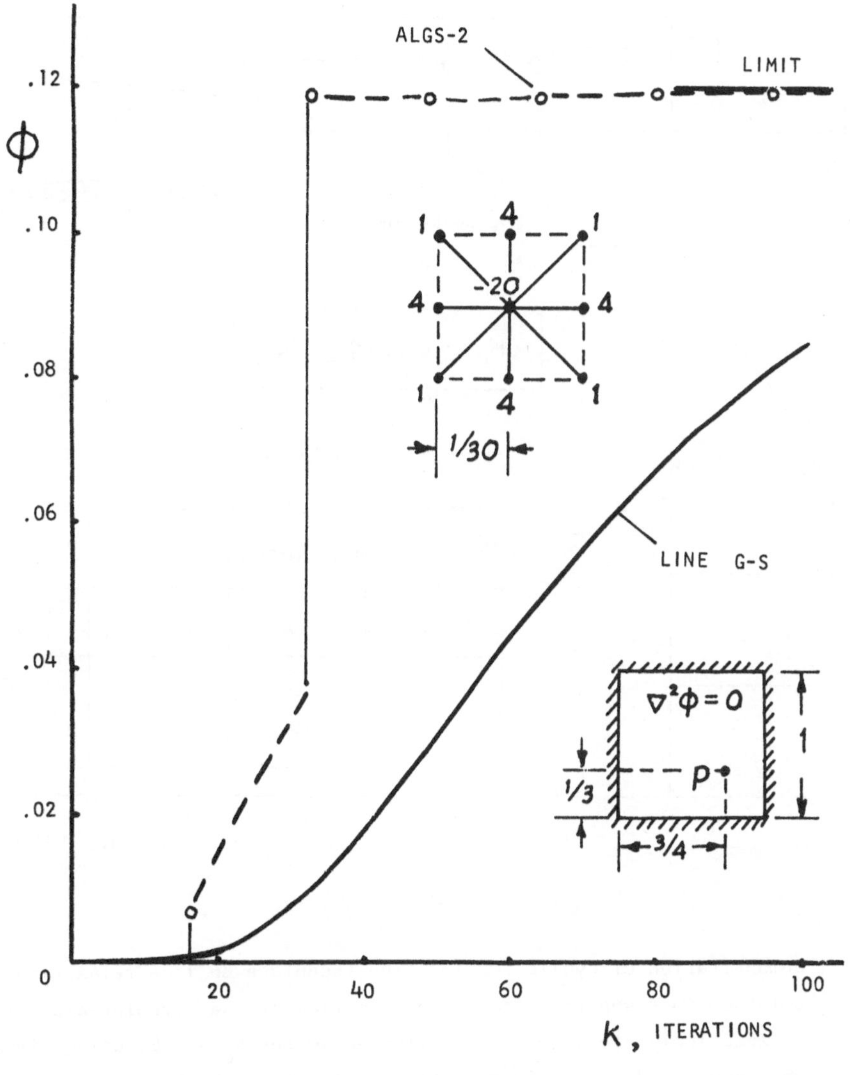

Figure 2. Comparison of unaccelerated and accelerated line SOR solutions to a model Dirichlet problem.

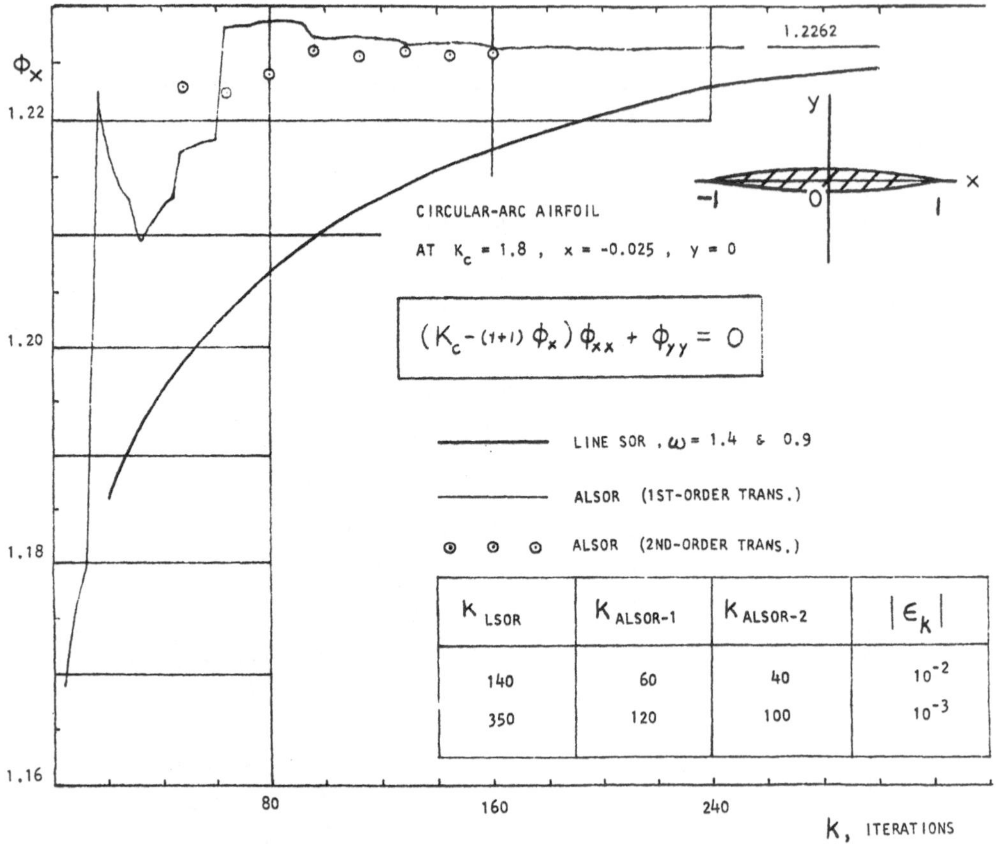

K_{LSOR}	$K_{ALSOR-1}$	$K_{ALSOR-2}$	$\lvert \epsilon_k \rvert$
140	60	40	10^{-2}
350	120	100	10^{-3}

(a)

Figure 3. Demonstration of cyclic acceleration technique on line relaxation solution to a supercritical transonic flow over a circular-arc airfoil at zero incidence with a similarity parameter $K_c = 1.8$, using first and second order transforms: (a) $\omega = 1.4$ and 0.9, and $\omega = 0.95$, uniform.

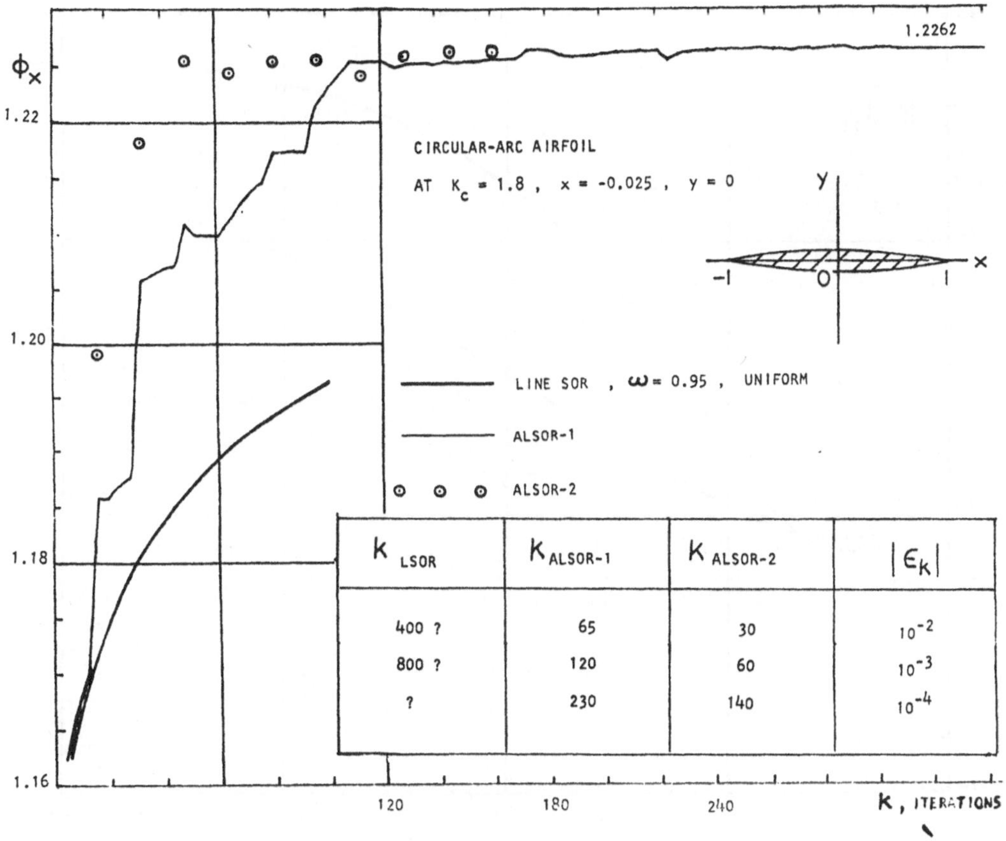

K_{LSOR}	$K_{ALSOR-1}$	$K_{ALSOR-2}$	$\lvert\epsilon_K\rvert$
400 ?	65	30	10^{-2}
800 ?	120	60	10^{-3}
?	230	140	10^{-4}

(b)

Figure 3 (continued)

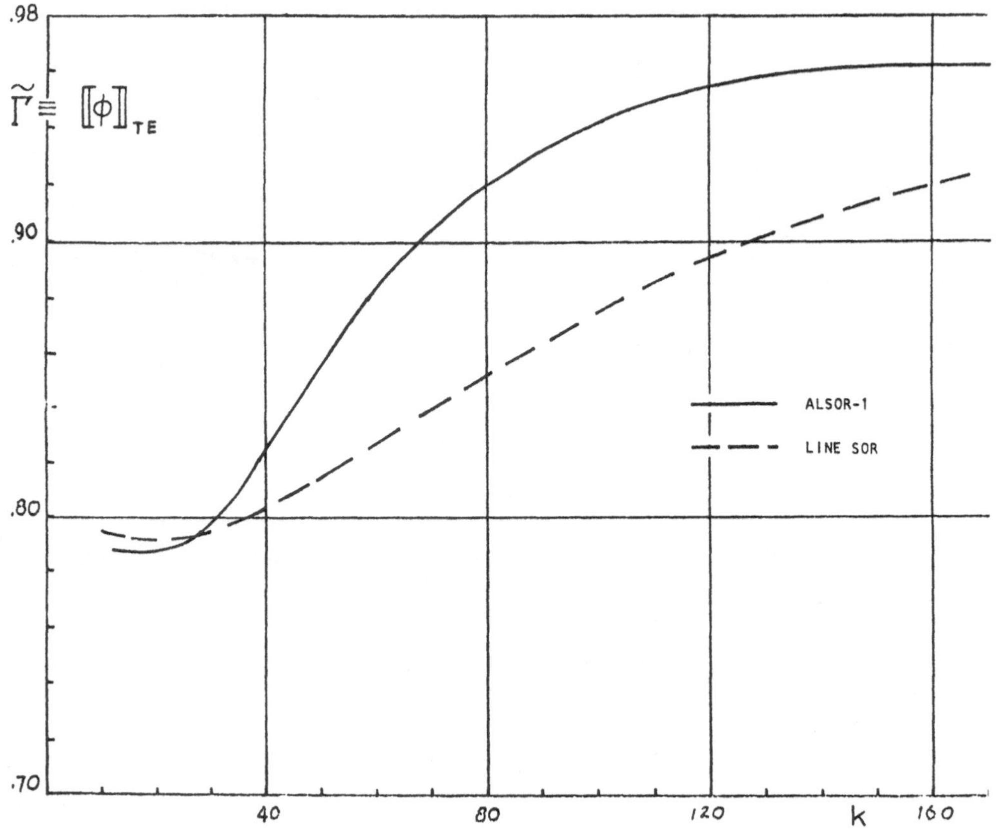

Figure 4. Test of cyclic acceleration technique on line SOR solution with ω = 1.8 and 0.8 to a supercritical transonic flow over a circular-arc airfoil at incidence with K_c = 1.8, α/τ = 0.1454, using first-order transform: Convergence history of the circulation.

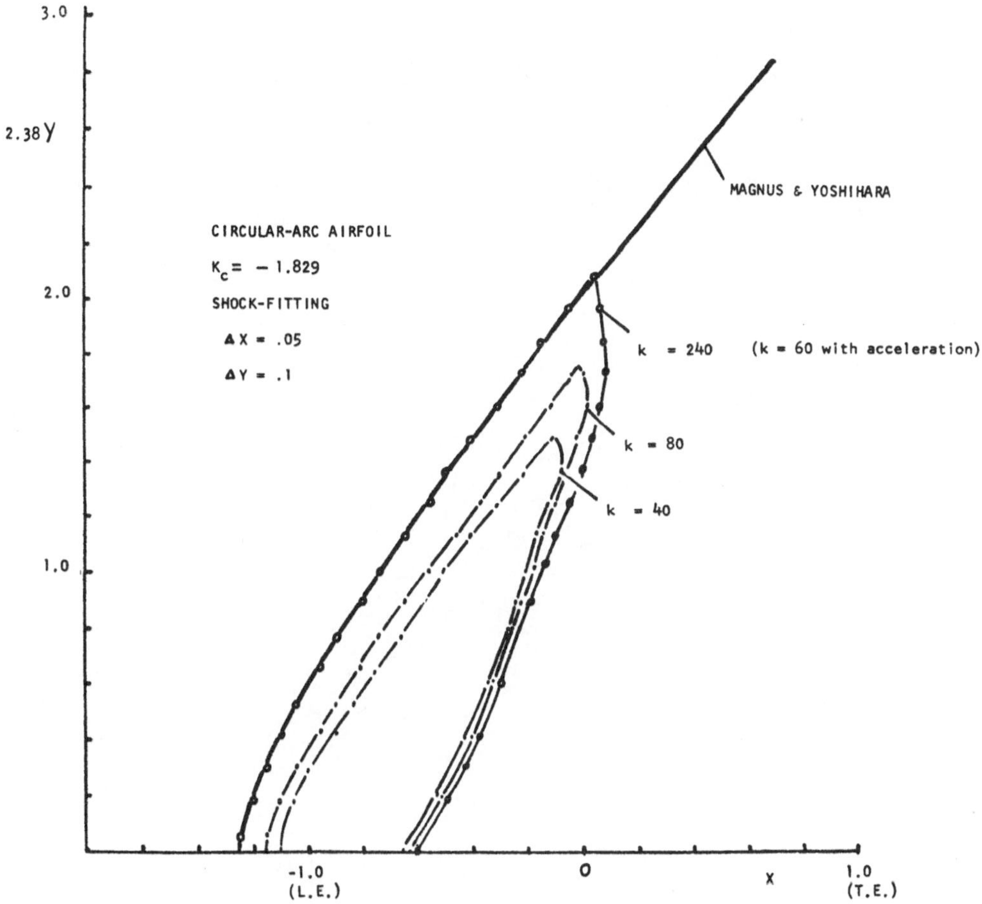

Figure 5. Application of cyclic acceleration technique to a line relaxation solution with shock fitting.

A Technique for Accelerating Iterative Convergence in Numerical Integration, with Application in Transonic Aerodynamics

E. DALE MARTIN
NASA Ames Research Center
Moffett Field, California, USA

Summary

A technique is described for the efficient numerical solution of nonlinear partial differential equations by rapid iteration. In particular, a special approach is described for applying the Aitken acceleration formula (a simple Padé approximant) for accelerating the iterative convergence. The method finds the most appropriate successive approximations, which are in a most nearly geometric sequence, for use in the Aitken formula. Simple examples are given to illustrate the use of the method. The method is then applied to the mixed elliptic-hyperbolic problem of steady, inviscid, transonic flow over an airfoil in a subsonic free stream.

1. Introduction

The numerical solutions of nonlinear partial differential equations such as those governing fluid flows frequently are obtained most efficiently by iterative methods. The rate of iterative convergence of the method chosen is an important consideration, and various means of accelerating the iterative convergence have been useful.

One popular device for accelerating convergence of a sequence of numbers such as provided by iteration is Aitken's extrapolation formula (or Δ^2 process) [1], whose use is described in most books on numerical methods [2] and which is identified [3–5] as a simple Padé approximant if the successive iterates are partial sums of a power series. Shanks [3] provided generalizations of Aitken's transformation and studied their use. In [6] Wynn gave a simple algorithm for rapid computation of one of the nonlinear transforms studied by Shanks, and later Wynn [7] discussed application of this acceleration technique to vector and matrix problems, including application to boundary-value and initial-value problems.

The present paper describes a special technique for applying the Aitken extrapolation formula for accelerating iterative convergence in the numerical solution of partial differential equations. The method was first introduced and used in [8] and then used in a modified form in [9] with additional results given in [9,10]. Although the application to be discussed is in a numerical finite-difference solution, the general method applies equally well, for example, to analytical solutions or to numerical solutions by finite-element methods. The use of the simple Aitken formula with three successive iterates is emphasized (even though the elegant ϵ-algorithm of Wynn with longer sequences could be used), because the eventual applications are expected to be those numerical problems requiring significant computer storage. The Aitken formula, using only three iterates, requires less storage than other forms of the ϵ-algorithm.

Often the use of the Aitken formula with iterates obtained arbitrarily by successive approximations does not lead to a significantly improved approximation. However, because Shanks [3] showed that the formula works best if the sequence is "nearly geometric," the present approach seeks to obtain successive iterates that are in a nearly geometric sequence. (Because of the work of Shanks in popularizing the Aitken formula and his valuable demonstration of the special applicability to "nearly geometric sequences," our past work has referred to the simple extrapolation formula as the "Aitken/Shanks formula.") The sequence of approximations can be most nearly geometric if obtained from a power-series construction. Therefore, the basis of the present approach is the construction of successive approximations derived from formal power-series expansions to obtain as closely as possible a nearly geometric sequence. The technique is based on the concepts of perturbation-series expansions (in the sense of Poincaré; see Bellman [11]). An artificial parameter is introduced in such a way as to obtain three problems to solve for terms of a nearly geometric series, for use in the Aitken/Shanks formula. Expansion in powers of an artificial parameter has also been considered by Genz [5] to develop a mathematical proof (unknown by the authors of [8] at that writing), but the central idea in the present approach is that the artificial-parameter expansions are used, in combination with an "artificially extended form" of the equations to be solved, as a device to determine most appropriate successive approximations. This technique produces the nearly geometric sequence of solutions, even in nonlinear problems. The previous application of the Aitken/Shanks transformations to acceleration of iterations in numerical integration by Wynn [7] used simple straightforward iterations. The results of such a procedure with use of only the simplest acceleration formula are described below for an example problem and are compared with the present method.

The present approach based on perturbation series requires that complete perturbation solutions be available on the entire computation field (or entire domain of the equations) at each iteration. This concept therefore adapts well to a finite-difference method using "direct elliptic solvers" [12–15] in the iterative procedure to determine the solution simultaneously at all points on the entire computation field (rather than in successive traverses over the field as in a point- or line-relaxation method). Such methods have been referred to as "semidirect" [8–10].

After several simple examples to illustrate the method, it is applied to the problem of inviscid flow over an airfoil in a subsonic free stream, including conditions for which the flow equations are of mixed type (elliptic in an outer region, with an embedded hyperbolic region and a shock wave). This transonic-aerodynamic-flow problem has also been treated by Hafez and Cheng [16] using the Aitken/Shanks acceleration formula, but in a quite different way, in combination with a line-relaxation method.

2. General Formulation of Method

Consider the general partial-differential or difference equation system and the accompanying boundary conditions represented by

$$LU - F(x) = NU \quad \text{in } R, \tag{2.1}$$

$$BU = G(x) \quad \text{on } B, \tag{2.2}$$

where $U = U(x)$ is a vector function of the position vector x, L is a separable, linear, elliptic differential or difference operator, $F(x)$ is a given vector function and N is a possibly nonlinear operator such that the operation NU is a vector of the same dimension as U and has components that may involve U, x, and derivatives of the components of U with respect to the components of x. Assume for simplicity that B is a linear operator. The boundary condition (2.2) is applied on B, which includes all appropriate boundary segments of the domain R. For illustration of this notation and of the method, simple one-dimensional examples are given in the next section. Examples treated in the earlier version of [8] included (i) the scalar Laplacian as L with a scalar, ψ, as U, and (ii) a Cauchy-Riemann operator matrix as L with two components of U, denoted as u and v. The right side of (2.1) can be complicated and can make the equation system hyperbolic or parabolic in some regions [8–10].

In the formulation of a problem to be solved, L and $F(x)$ are chosen judiciously and may be the result of "scaling and shifting" transformations [17,9] for increasing the rate of iterative convergence or of addition of terms [9,10] for stabilizing iterations. For treatment with additional terms, an extended Cauchy-Riemann solver for use in present calculations has been described in [18].

In the methods to be discussed for the iterative solution of eqs. (2.1) and (2.2), suppose $U_1(x)$, $U_2(x)$, and $U_3(x)$ are successive approximations to $U(x)$ in R. Let $u(x)$ and $u_n(x)$ be respectively each a single scalar component of the vectors $U(x)$ and $U_n(x)$ $(n = 1,2,3)$. Then one form of the Aitken/Shanks extrapolation formula [1,3] for an improved approximation $u^*(x)$ to $u(x)$ is

$$u^*(x) = \frac{u_1 u_3 - u_2^2}{u_1 - 2u_2 + u_3} \tag{2.3}$$

Application of the formula in this way to individual components of U at each x separately is referred to by Wynn [7] as use of a "primitive inverse" of the ϵ-algorithm. Wynn concludes that use of the primitive inverse is competitive with use of other more complicated inverses. The work of Hafez and Cheng [16] considers coupling of the matrix elements in the numerical solution, which is related to the more complex inverses of the ϵ-algorithm.

2.1 Artificially extended equation. For obtaining power-series solutions to (2.1) and (2.2) that are most appropriate for use in the Aitken/Shanks extrapolation formula, it has been found convenient to artificially extend eq. (2.1) by inserting both an artificial parameter ϵ and an "initial approximation," $U_0(x)$, to $U(x)$ as follows. Let

$$LU - F(x) = (1-\epsilon)NU_0 + \epsilon\, NU \quad \text{in } R \tag{2.4}$$

along with condition (2.2). Note that the solution U to (2.4) with (2.2) depends on ϵ (as well as on the specified function $U_0(x)$): $U = U(x,\epsilon)$. However, at $\epsilon = 1$, the solution to (2.4) with (2.2) is the same as the solution to the original equations (2.1) with (2.2). Furthermore, if $U_0(x)$ is close to the solution $U(x)$, then (2.4) is nearly the same as (2.1) and the solutions then are nearly the same. Thus, either of the conditions $\epsilon = 1$ or $U_0 = U$ makes (2.4) the same as (2.1). Both of these facts can be used to advantage in the methods to be discussed.

2.2 Method 1. The simplest iteration scheme is a straightforward method of successive approximations. Although this method can be combined with use of a relaxation parameter (see [8,9]), for simplicity here we omit that useful device. If we let $\epsilon = 0$ in (2.4) and define $U_0(x)$ as a previous iteration, we obtain the following equations for the iterative solution denoted as Method 1(a):

$$LU_n - F = NU_{n-1} \quad \text{in } R, \tag{2.5a}$$

$$BU_n = G(x) \quad \text{on } B, \tag{2.5b}$$

where subscript n denotes iteration number.

If, as is frequently done, the Aitken/Shanks formula is used to attempt to accelerate the convergence of the iteration, we denote as Method 1(b) the solution of (2.5) for three successive iterates and substitution of the results for one component of each U_n into (2.3). (This designation of Method 1(b) is useful for a comparison in an example problem below.)

2.3 Method 2. The new approach for applying the Aitken/Shanks formula, first introduced and used in [8] and in a modified form in [9], is referred to as Method 2. The two versions are called, respectively, Methods 2(a) and 2(b) for later convenience.

Consider the solution to (2.4) with condition (2.2). The solution evaluated at $\epsilon = 1$ is a solution to (2.1) with (2.2). The specified $U_0(x)$ can be used as an initial approximation to U. For obtaining a most nearly geometric sequence of approximations, assume that

$$U(x,\epsilon) \sim U_1'(x) + \epsilon U_2'(x) + \epsilon^2 U_3'(x) + \dots . \tag{2.6}$$

Successive approximations to $U(x)$ are then defined by n-term truncations of the series (2.6):

$$U_n = \sum_{i=1}^{n} \epsilon^{i-1} U_i'(x) \tag{2.7}$$

Although (2.6) is equivalent to a Taylor series or asymptotic series expansion about $\epsilon = 0$, its convergence or lack of convergence at $\epsilon = 1$ is not of particular significance for applicability of eq. (2.3) (see [3]). If the series (2.6) is substituted into the problem of eq. (2.4) and condition (2.2) and coefficients of powers of ϵ are collected, one obtains equations to solve for the U_n':

$$LU_1' - F = NU_0 \quad \text{in } R; \qquad BU_1' = G(x) \text{ on } B; \tag{2.8a}$$

$$LU_2' = NU_1' - NU_0 \quad \text{in } R; \qquad BU_2' = 0 \text{ on } B; \tag{2.8b}$$

$$LU_3' = N_2'\{U_2',U_1'\} \quad \text{in } R; \qquad BU_3' = 0 \text{ on } B; \tag{2.8c}$$

in which N_2' is defined by the perturbation expansion

$$NU = NU_1' + \epsilon N_2' \{U_2',U_1'\} + O(\epsilon^2). \tag{2.9}$$

With the definitions (2.7) and

$$N_2 \{U_2,U_1\} \equiv NU_1' + \epsilon N_2' \{U_2',U_1'\} \tag{2.10}$$

one can also solve the following equations for the successive approximations, U_n:

$$LU_1 - F = NU_0 \quad \text{in } R; \qquad BU_1 = G(x) \text{ on } B \tag{2.11a}$$

$$LU_2 - F = NU_1 \quad \text{in } R; \qquad BU_2 = G(x) \text{ on } B \tag{2.11b}$$

$$LU_3 - F = N_2 \{U_2,U_1\} \quad \text{in } R; \qquad BU_3 = G(x) \text{ on } B \tag{2.11c}$$

in which it has been assumed that $\epsilon = 1$. Note that if the right side of eq. (2.1) is linear in $U(x)$, then the problems for the successive U_n in eqs. (2.11) are the same as (2.5) for Method 1.

We denote as Method 2(a) the solution of eqs. (2.11) for three successive iterates and substitution of the results for one component of each U_n into (2.3) to obtain an improved approximation. (If NU is linear in U, this is the same as Method 1(b)). Note that when the solution is near to convergence at any x, significant errors will be introduced by the loss of significant figures in applying eq. (2.3).

An alternative procedure (denoted as Method 2(b)) that eliminates the difficulty near convergence is to replace eq. (2.3) by the equivalent expression (at $\epsilon = 1$):

$$u^*(x) = u_1' - \frac{(u_2')^2}{u_3' - u_2'}, \tag{2.12}$$

where each $u_n'(x)$ is a single component of the vector $U_n'(x)$. That is, eqs. (2.8) are solved for $U_n'(x)$, and (2.12) is used for extrapolation.

In a numerical solution, $u^*(x)$ can be used as the next $u_0(x)$ in a repetition of the sequence.

3. Example Problems and Comparison of Methods

This section gives simple analytical one-dimensional examples for illustration and comparison of the methods.

3.1 Example 1. Consider the nonlinear problem

$$(d/dx + 1)u = (1/2)u^2 \quad \text{in } 0 \leqslant x < \infty, \tag{3.1a}$$

$$u(0) = 1. \tag{3.1b}$$

The iterative solution by Method 1 is found from

$$(d/dx + 1)u_n = (1/2)u_{n-1}^2 \quad , \quad u_n(0) = 1 . \tag{3.2}$$

The analytical solutions for $n = 1,2,3$ (assuming $u_o = 0$) are:

$$u_1(x) = e^{-x} , \tag{3.3a}$$
$$u_2(x) = e^{-x}\,[1 + p(x)] , \tag{3.3b}$$
$$u_3(x) = e^{-x}\,[1 + p(x) + p^2(x) + \frac{1}{3}\,p^3(x)] , \tag{3.3c}$$

where

$$p(x) = (1/2)\,(1-e^{-x}) . \tag{3.4}$$

For Method 2, the artificially extended equation is:

$$(d/dx + 1)u = (1-\epsilon)\,(1/2)u_o^2 + \epsilon(1/2)u^2 , \tag{3.5a}$$
$$u(0) = 1 . \tag{3.5b}$$

Substitution of

$$u = u_1'(x) + \epsilon u_2'(x) + \epsilon^2 u_3'(x) + \ldots \tag{3.6}$$

into (3.5) leads to

$$(d/dx + 1)u_1' = (1/2)u_o^2 , \qquad u_1'(0) = 1 , \tag{3.7a}$$
$$(d/dx + 1)u_2' = (1/2)\,[(u_1')^2 - u_o^2] , \qquad u_2'(0) = 0 , \tag{3.7b}$$
$$(d/dx + 1)u_3' = u_1'u_2' , \qquad u_3'(0) = 0 , \tag{3.7c}$$

or equivalently, with $\epsilon = 1$ and eq. (2.7),

$$(d/dx + 1)u_1 = (1/2)u_o^2 , \qquad u_1(0) = 1 , \tag{3.8a}$$
$$(d/dx + 1)u_2 = (1/2)u_1^2 , \qquad u_2(0) = 1 , \tag{3.8b}$$
$$(d/dx + 1)u_3 = (1/2)u_2^2 - (1/2)(u_1-u_2)^2 , u_3(0) = 1 . \tag{3.8c}$$

The analytical solutions to (3.7) with $u_o = 0$ are:

$$u_n'(x) = e^{-x}[p(x)]^{n-1} , \tag{3.9}$$

where $p(x)$ is given by (3.4) and where the solutions u_n to (3.8) are given by (2.7). Evaluations of these solutions at $x = 1$ and applications of the appropriate forms of the Aitken/Shanks formula are given in Table 1. The results for the extrapolated solution u^* may be compared with the exact solution to (3.1),

$$u(x) = 2(1+e^x)^{-1} , \tag{3.10}$$

Table 1. Results of Example 1 at $x = 1$ ($u_0 = 0$)

METHOD:	1(b)	2(a)	2(b)
EQUATIONS:	(3.2) & (2.3)	(3.8) & (2.3)	(3.7) & (2.12)
n	$u_n(1)$	$u_n(1)$	$u_n'(1)$
1	0.3678794412	0.3678794412	0.3678794412
2	.4841515202	.4841515202	.2325441579
3	.5247721376	.5209005060	.1469959430
$u^*(1)$ =	.546583145	.537882842	.5378828426
Exact $u(1)$ =	.5378828428	.5378828428	.5378828428

evaluated at $x = 1$: $u(1) = 0.5378828428$ to ten significant figures. We note first that the extrapolated solution u^* by Method 1(b) is somewhat closer to the exact value than u_3, but not significantly closer. We note further that the third approximation, u_3, by Method 2(a) is not as good an approximation as u_3 in Method 1, but that the extrapolated solutions by Methods 2(a) and 2(b) are exact except for loss of 1 or 2 significant figures. (Method 2(a) is less exact because of loss of significant figures in (2.3).) The striking accuracy of Method 2 in this example occurs because the sequence of solutions produced by Method 2 is precisely geometric, i.e. $u_{n+1}'/u_n' =$ constant for all n at a given x. The difference from Method 1 is seen by comparing eqs. (3.2) with (3.8), in which (3.8c) has an additional term that produces the geometric sequence.

3.2 Example 2. Consider next an example which is linear (so that Method 2(a) would give the same results as Method 1(b)), but for which the iterative sequence is "nearly geometric." Let us use Method 2(b) for this example (eqs. (2.8) with (2.6), (2.7), and (2.12)).

The problem is

$$\frac{du}{dx} - 2 = -x\frac{du}{dx} - 2u \quad \text{in } 0 \leqslant x < \infty \ , \quad u(0) = 0 \ , \tag{3.11}$$

which is written in this way in analogy to more complex problems in which one may put a very simple operator on the left and the rest of the terms on the right for iteration. (One can also shift the term $2u$ to the left side, with very similar results.) The artificially extended equation is

$$\left.\begin{aligned} \frac{du}{dx} - 2 = (1-\epsilon)\left(-x\frac{du_0}{dx} - 2u_0\right) + \epsilon\left(-x\frac{du}{dx} - 2u\right) \quad \text{in } 0 \leqslant x < \infty \\ u(0) = 0 \end{aligned}\right\} \tag{3.12}$$

Substitution of (3.6) leads to (with $u_0 = 0$):

$$du_1'/dx - 2 = 0\,, \qquad u_1'(0) = 0\,, \tag{3.13a}$$

$$du_2'/dx = -x\, du_1'/dx - 2u_1'\,, \qquad u_2'(0) = 0\,, \tag{3.13b}$$

$$du_3'/dx = -x\, du_2'/dx - 2u_2'\,, \qquad u_3'(0) = 0\,. \tag{3.13c}$$

The analytical solutions are

$$u_n'(x) = (-1)^{n+1}(n+1)x^n \tag{3.14}$$

and the successive approximations are given by (2.7). The sequence (3.14) is not geometric, but since $\ell im_{n\to\infty} [u_{n+1}'(x)/u_n'(x)]$ exists at given x, the sequence is "nearly geometric" [3]. Evaluation of the solutions (3.14) at $x = 0.5$ gives $(u_1', u_2', u_3') = (1.00, -.75, .50)$ so that the successive approximations are $(u_1, u_2, u_3) = (1.00, .25, .75)$. Substitution of the u_n' into (2.12) gives $u^*(.5) = 0.55$, which compares well with the exact solution to (3.11),

$$u(x) = (2x + x^2)(1 + x)^{-2}\,, \tag{3.15}$$

from which $u(0.5) = 5/9 = 0.555555\ldots$.

4. Transonic Flow Over an Airfoil

For application of the methods described above, consider two-dimensional, steady, inviscid flow over a thin symmetrical parabolic-arc airfoil in a subsonic free stream. At high subsonic Mach numbers, part of the flow can be supersonic, so we consider the transonic small-disturbance equations, which are nonlinear elliptic partial differential equations in subsonic regions and hyperbolic equations in supersonic regions. Transition of the velocity field from a subsonic region to the embedded supersonic zone is smooth, but transition from the supersonic to subsonic region is usually discontinuous, through a shock wave. The improved finite-difference method of Murman and Cole [19–22] captures the shock waves (in a fully conservative way) but spreads the rapid transition over several mesh points.

In [8] a semidirect finite-difference method, based on the use of a fast direct Cauchy-Riemann solver [15], was applied to solving the equivalent of Murman's transonic finite-difference equations [21] iteratively for the perturbation velocities, u and v. (The iteration procedure has been formulated in such a way that at nonelliptic points terms on the right side of the difference equations cancel out the elliptic character of the left side when the iterated solution converges.) Both Methods 1(a) and 2(a) described above worked well for subcritical and for slightly supercritical (local Mach number > 1) flows, except that Method 2(a) could be used only before any part of the solution was nearly converged. In [9] the method was extended to strongly supercritical flow by the addition of stabilizing terms to the difference equations and to the Cauchy-Riemann solver [18]. Also introduced in [9] was the method version denoted here as Method 2(b), which can be used when the solution is nearly converged. In smooth subsonic flows the acceleration technique is effectively used repeatedly. However, in transonic flows with strong shock waves, the acceleration technique is not helpful at the beginning of the iteration when the shock wave and its location are not well defined. Therefore in [9] it was considered desirable to use the straightforward iteration Method 1(a) until the maximum residual is reasonably small, so that the supersonic region is nearly defined, and then use Method 2(b) to extrapolate three iterates to a final solution. A fully conservative second-order-accurate formulation has been introduced in [10], and so a fomulation that includes either Murman's fully-conservative first-order-accurate formulation or the second-order formulation will be used here.

4.1 Governing equations and boundary conditions. Let the dimensionless X and Y axes be respectively along and normal to the airfoil chord, the free-stream Mach number be $M_\infty < 1$, and the dimensionless velocity components in the X and Y directions be U,V. One may then define perturbation velocity components u,v through a Prandtl-Glauert transformation with $\beta \equiv (1 - M_\infty^2)^{1/2}$:

$$U = 1 + (\tau/\beta)u\,, \quad V = \tau v\,, \quad Y = y/\beta\,, \quad X = x\,, \tag{4.1}$$

which amounts to shifting and scaling of certain terms (cf. [17, 8–10], so that the transonic small disturbance equations take the form

$$f_x + g_y = 0\,, \quad u_y - v_x = 0 \tag{4.2a,b}$$

where

$$f = f(u) = u - au^2\,, \quad g = g(v) = v\,, \tag{4.3a,b}$$

$$a = \tau(\gamma+1)M_\infty^2/2\beta^3\,, \tag{4.4}$$

in which a is a transonic similarity parameter and τ is an airfoil thickness ratio. Eqs. (4.2) are often written in terms of a perturbation velocity potential ϕ defined by $u = \phi_x$, $v = \phi_y$, and all the developments to be described apply as well to that potential equation.

The equation system (4.2) is elliptic, parabolic, or hyperbolic depending on whether $u - u_{CR}$ is negative, zero, or positive, where the transformed critical velocity is $u_{CR} = 1/2a$. The corresponding pressure coefficient is $C_p = -2(\tau/\beta)u$.

The linearized surface boundary condition for the symmetrical parabolic-arc airfoil, whose upper surface is given by $Y_b(x) = \tau F(x) = \tau(0.5 - 2x^2)$ in $-.5 \leqslant x \leqslant .5$ (with $F(x) = 0$ in $|x| > .5$), and the conditions at infinity are

$$v(x,0^+) = F'(x)\,, \tag{4.5a}$$

$$u,v \to 0 \quad \text{as} \quad x^2 + y^2 \to \infty\,. \tag{4.5b}$$

Eq. (4.2a) is written in a "conservation-law" (or divergence) form, in terms of flux components f and g. Therefore discretized forms of (4.2a), for numerical solution, can represent in a fully conservative way either that differential equation or the corresponding integral form. These discretized forms can thus be formulated correctly to represent transitions between elliptic and hyperbolic regions [21,10].

Since only the term f_x in the system (4.2) determines the type of point (depending on the local value of u), one can write the general type-dependent difference equations in the form:

$$(f_x)_T + (g_y)_C = 0\,, \quad (u_y)_C - (v_x)_C = 0\,, \tag{4.6a,b}$$

where subscript C indicates a central-differenced representation of a derivative and subscript T, which indicates type-dependent differencing, may be replaced by E, H, P, or S at points defined respectively as elliptic, hyperbolic, parabolic, or shock points [21,10]. At all points where the difference equations are clearly elliptic or hyperbolic, subscripts E or H are used. Transition points from elliptic to hyperbolic (progressing downstream from left to right) are P points, and transitions from hyperbolic to elliptic are S points.

For defining the finite-difference operators, Fig. 1 shows a staggered u,v mesh, with the shaded area indicating a mesh cell for eq. (4.2a). The center of a mesh cell is the point at which ϕ would be defined on a conventional mesh and is the point that is designated E, H, P, or S. The indices j and k indicate respectively the x and y directions. Second-order-accurate central differences are

$$\left.\begin{aligned} (u_x)_C &= (u_{j,k} - u_{j-1,k})/\Delta x\,, & (v_y)_C &= (v_{j,k} - v_{j,k-1})/\Delta y\,, \\ (u_y)_C &= (u_{j,k+1} - u_{j,k})/\Delta y\,, & (v_x)_C &= (v_{j+1,k} - v_{j,k})/\Delta x\,. \end{aligned}\right\} \tag{4.7}$$

In general, $(f_x)_T$ is represented by

$$\Delta x (f_x)_T \equiv \Delta f_{j,k} = (f_G)_{j,k} - (f_G)_{j-1,k} \tag{4.8a}$$

where

$$(f_G)_{j,k} = f((u_G)_{j,k}) = (u_G)_{j,k} - a(u_G^2)_{j,k} \tag{4.8b}$$

and where u_G is either a "hyperbolic form" u_H or an "elliptic form" u_E. With (i) the definition (4.8) for the difference operator $(f_x)_T$, (ii) a condition to determine whether each $(u_G)_{j,k}$ is represented by u_E or u_H, and (iii) specifications of u_E and u_H to obtain the finite differences (4.8a) to the order of

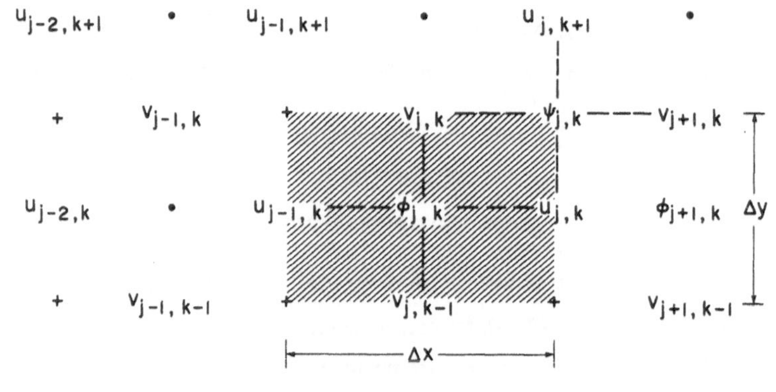

Fig. 1 – Differencing mesh and mesh cell.

accuracy desired, all four type-dependent operators are obtained. As derived in [10], (i) the second-order-accurate elliptic operator, (ii) either the first-order or second-order-accurate hyperbolic operator, and (iii) the corresponding parabolic-point and shock-point operators are all produced in (4.6) with (4.7) and (4.8) by the following relationship:

$$(u_G)_{j,k} = (1-\sigma_{j,k})u_{j,k} + \sigma_{j,k}[\lambda u_{j-1,k} + (1-\lambda)u_{j-2,k}] , \tag{4.9}$$

where

$$\left.\begin{aligned} \sigma_{j,k} &= 0 \text{ (and } u_G = u_E) \quad \text{if} \quad \tilde{u}_{j,k} < u_{CR} , \\ &= 1 \text{ (and } u_G = u_H) \quad \text{if} \quad \tilde{u}_{j,k} > u_{CR} , \end{aligned}\right\} \tag{4.10}$$

$$\tilde{u}_{j,k} = (u_{j,k} + \delta\, u_{j-1,k})/(1+\delta) , \tag{4.11}$$

and where $\lambda = 1$ for the first-order-accurate hyperbolic operator, $\lambda = 2$ for the second-order-accurate hyperbolic operator, and δ is a parameter that may be varied from 0 to ∞ but is derived as unity for Murman's first-order-accurate operators [21]. As an example to illustrate, suppose $\lambda = 1$, $\delta = 1$, and $\tilde{u}_{j,k} < u_{CR}$ and $\tilde{u}_{j-1,k} > u_{CR}$. Then for the shaded mesh cell in Fig. 1, eqs. (4.8) – (4.10) give

$$\Delta f_{j,k} = u_{j,k} - u_{j-2,k} - a(u_{j,k}^2 - u_{j-2,k}^2)$$

which is equivalent to Murman's [21] first-order shock-point operator. In a similar way the Krupp-Murman first-order parabolic operator [20] is also obtained. Both the first- and second-order-accurate hyperbolic operators given by (4.8) – (4.10) with $\lambda = 1$ and $\lambda = 2$ are equivalent to upwind difference operators originally proposed by Murman and Cole [19]; the fully conservative second-order P and S operators were introduced in [10]. Analysis of all these E, H, P, and S operators [10] has verified their consistency, accuracy, and stability in the examples computed.

Because of the slow iterative convergence of the second-order-accurate iterative method to be described, two methods of adding artificial viscosity have been proposed and used [10]. Both leave the scheme fully conservative and formally second-order-accurate.

The boundary conditions for the finite-difference equations (4.6) are the same as (4.5) but with (4.5b) replaced by a far-field condition on an outer rectangular boundary B:

$$u_{j,k} = u_B(x,y) \quad \text{or} \quad v_{j,k} = v_B(x,y) \quad \text{on B} \tag{4.12}$$

where, for example, u_B and v_B are given by a Prandtl-Glauert solution (see [15,8,9]).

For solution of eqs. (4.6) with (4.7) through (4.11) and with conditions (4.12) by the semidirect methods, one must rearrange the equations so that the left side is an appropriate elliptic operator and provides a stable iteration scheme. One first adds $(u_x)_C - (f_x)_T$ to both sides of (4.6a) to obtain

$$(u_x)_C + (v_y)_C = (u_x)_C - (f_x)_T , \tag{4.13a}$$

$$(u_y)_C - (v_x)_C = 0 . \tag{4.13b}$$

This set contains a central-differenced elliptic operator on the left side regardless of the local type of the equations. The nonlinear type-dependent term has been shifted to the right side where, in an iterative procedure, it can be computed from a previous iteration. Although the iteration of these equations [8] converged well for subsonic and slightly supercritical flow, it was found [9,10] that terms with parameters multiplying $u_{j,k}$ and $u_{j-1,k}$ needed to be added to both sides of (4.13a) to produce iterative convergence at higher Mach numbers. A more specific form of the difference equations, in which the second-order-accurate relations (4.7) have been substituted, is

$$D_{j,k}(u,v) = R_{j,k}(u) , \quad E_{j,k}(u,v) = 0 , \tag{4.14a,b}$$

in which

$$D_{j,k}(u,v) \equiv (1-\alpha_1)u_{j,k} - (1+\alpha_2)u_{j-1,k} + \mu^{-1}(v_{j,k} - v_{j,k-1}) , \tag{4.15a}$$

$$E_{j,k}(u,v) \equiv (u_{j,k+1} - u_{j,k}) - \mu(v_{j+1,k} - v_{j,k}) , \tag{4.15b}$$

$$R_{j,k}(u) \equiv (1-\alpha_1)u_{j,k} - (1+\alpha_2)u_{j-1,k} - \Delta f_{j,k} , \tag{4.15c}$$

and where $\Delta f_{j,k}$ is defined by eqs. (4.8) – (4.11) and $\mu \equiv \Delta y/\Delta x$. The formal order of accuracy of eqs. (4.14) depends on the value of λ used in (4.9).

4.2 Equations for Method 1(a). As described in section 2.2 above, the straightforward iteration Method 1(a) for eqs. (4.14) is simply

$$D_{j,k}(u_n,v_n) = R_{j,k}(u_{n-1}) , \quad E_{j,k}(u_n,v_n) = 0 . \tag{4.16a,b}$$

For determining each $\sigma_{j,k}$ in (4.10), eq. (4.11) uses u_{n-1}. The presence of $\alpha_1 u_{j,k}$ and $\alpha_2 u_{j-1,k}$ on both sides of eq. (4.16a) allows the interpretation and treatment of these terms as an off-centered time derivative, $\partial u/\partial t$, multiplied by a constant. When the solution converges, these terms cancel out. The semidirect Method 1(a) proceeds by solving the left side of (4.16) in terms of the known right side by an "extended Cauchy-Riemann" solver [18] for u_n and v_n at all points simultaneously. The iteration with α_1 or $\alpha_2 \neq 0$ needs a reasonable (but very roughly approximate) initial approximation (u_0), such as a Prandtl-Glauert solution. Ref. [10] gives variable specifications of α_2 for best convergence.

The boundary conditions on (4.16) are

$$v_n(x,0^+) = F'(x) , \tag{4.17a}$$

$$u_n = u_B \quad \text{or} \quad v_n = v_B \quad \text{on } B . \tag{4.17b}$$

4.3 Equations for Method 2(b). The artificially extended form, (2.4), of eqs. (4.14) is

$$D_{j,k}(u,v) = (1-\epsilon)R_{j,k}(u_0) + \epsilon R_{j,k}(u) , \tag{4.18a}$$

$$E_{j,k}(u,v) = 0 . \tag{4.18b}$$

For Method 2(b) assume that

$$u(x,y,\epsilon) = u_1'(x,y) + \epsilon u_2'(x,y) + \epsilon^2 u_3'(x,y) + \dots , \qquad (4.19a)$$

$$v(x,y,\epsilon) = v_1'(x,y) + \epsilon v_2'(x,y) + \epsilon^2 v_3'(x,y) + \dots . \qquad (4.19b)$$

The successive approximations are then (for $n = 1,2,3 \ldots$)

$$u_n = \sum_{i=1}^{n} \epsilon^{i-1} u_i'(x,y) , \quad v_n = \sum_{i=1}^{n} \epsilon^{i-1} v_i'(x,y) . \qquad (4.20)$$

Substitution of (4.19) into (4.18) leads to

$$D_{j,k}(u_n', v_n') = R_{n-1} , \quad E_{j,k}(u_n', v_n') = 0 , \qquad (4.21)$$

where:

$$R_0 = R_{j,k}(u_0) \qquad (4.22a)$$

$$R_1 = R_{j,k}(u_1') - R_{j,k}(u_0) \qquad (4.22b)$$

(with u_0 being used in (4.11) in determining $\sigma_{j,k}$ for use in $R_{j,k}(u_1')$) and

$$R_2 = (1-\alpha_1)(u_2')_{j,k} - (1+\alpha_2)(u_2')_{j,k} - (\Delta f_2)_{j,k} , \qquad (4.22c)$$

$$\Delta f_2 = (f_2)_{j,k} - (f_2)_{j-1,k} , \qquad (4.23a)$$

$$\begin{aligned}(f_2)_{j,k} = {} & (1-\sigma_{j,k}) \, [(u_2')_{j,k} - 2a(u_1' u_2')_{j,k}] \\ & + \sigma_{j,k} \left\{ [\lambda(u_2')_{j-1,k} + (1-\lambda)(u_2')_{j-2,k}] \right. \\ & \left. - 2a[\lambda(u_1')_{j-1,k} + (1-\lambda)(u_1')_{j-2,k}] \, [\lambda(u_2')_{j-1,k} + (1-\lambda)(u_2')_{j-2,k}] \right\} . \end{aligned} \qquad (4.23b)$$

The boundary conditions are:

$$v_1'(x,0^+) = F'(x) ; \quad v_n'(x,0^+) = 0 \;\; (n = 2,3) ; \qquad (4.24a)$$

$$u_1' = u_B \;\; \text{or} \;\; v_1' = v_B \quad \text{on } B ; \qquad (4.24b)$$

$$u_n' = 0 \;\; \text{or} \;\; v_n' = 0 \quad \text{on } B \;\; (n = 2,3) . \qquad (4.24c)$$

With some reasonable approximation for $(u_0)_{j,k}$, such as a nearly converged solution by Method 1(a), eqs. (4.21), with $n = 1,2,3$, give three successive approximations u_1', u_2', u_3' at each j,k to use in (2.12) to obtain an extrapolated solution.

4.4 Results and discussion. A research computer program written to solve the transonic small disturbance equations by the methods described above for a biconvex airfoil at zero incidence, includes the option of switching after some iterations by Method 1(a) to the extrapolation technique, Method 2(a). A conversational version of the program, for interacting with the program, was run on an IBM 360/67 computer, and computing times were measured on a Control Data 7600 computer.

Pressure distributions have been computed for a range of subsonic and transonic Mach numbers from both first- and second-order-accurate formulations. Examples by Method 1(a) are shown on Fig. 2 for a thickness ratio of 10 percent and $M_\infty = 0.825$. For this calculation the boundaries were at one-half chord upstream and downstream of the airfoil edges and at 3.5 chords above the airfoil. The results computed on a 39X32 uniform mesh compare well with a line-relaxation program [22], which uses a variable and finer mesh. On a very coarse (19X32) mesh, with only 10 mesh intervals on the airfoil chord, the first-order-accurate results, of course, are not good. The shock is badly smeared, and an anomalous jump behind the sonic point that is characteristic of the first-order P operator is exaggerated on the coarse mesh. However, the second-order-accurate results are very smooth through the sonic point and are surprisingly accurate.

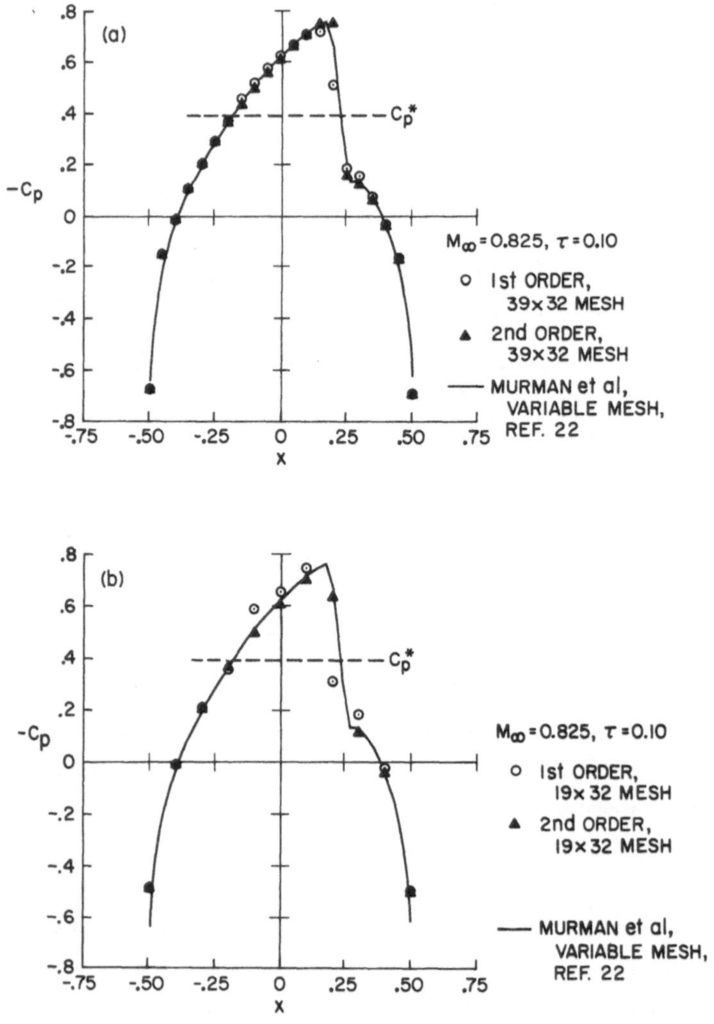

Fig. 2 – Pressure on a thin biconvex airfoil.

Figure 3 shows an interesting effect of switching to Method 2 before the iteration has converged enough, when the types of all points are not yet quite the same as the final types. Method 1(a) was used for nine iterations; then Method 2(b) was used to obtain the three successive terms at each point and the extrapolated solution shown in Fig. 3. A property of the Aitken /Shanks extrapolation as used in Method 2 is that all the significant figures of the three successive approximations at any point contain information about the exact solution, even though those successive approximations themselves are not very close to the exact solution (see example problems above in section 3). It thus appears possible in Fig. 3 that this procedure may be picking up the fact that the exact solution to the equations (or the solution on a very fine mesh) has the well-known logarithmic singularity just behind the shock, even though the converged solution on the coarse mesh smears over this singularity. Even the finer mesh used by the program in [22] was not fine enough to pick up the singularity, partly because that point apparently occurs between the mesh points for this case. This phenomenon illustrated in Fig. 3 is not an isolated case but is a typical occurrence in Method 2. It may be that the numerical solution in Fig. 3 is as good as representation of the exact solution to the equations as is the fully converged solution in Fig. 2(a) (circles).

The most significant property of the semidirect method is the relatively short computing time required. On the 39X32 mesh, the time per iteration was measured as 40 milliseconds in a very inefficiently coded program, but for various reasons discussed in [10] it is expected to be reduced to 20 ms. (The direct solver requires only 14 ms) The subcritical cases were sufficiently converged in 3 iterations or less, and a slightly supercritical case (first-order-accurate, using Method 2) required 6 iterations. The

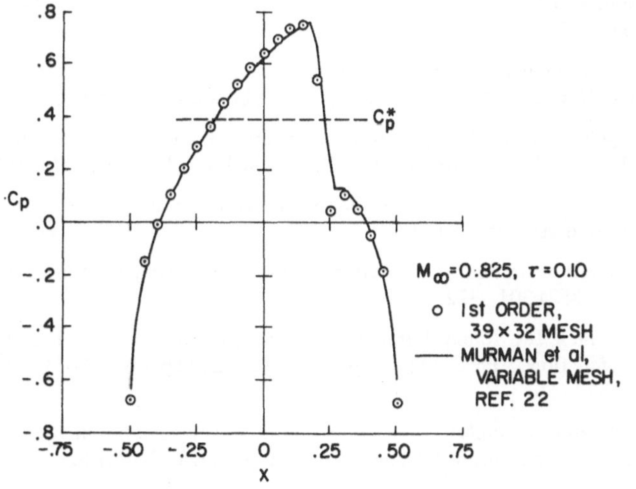

Fig. 3 – Pressure distribution resulting from Aitken/Shanks extrapolation (Method 2(b)) before iterative convergence.

first-order-accurate case shown in Fig. 2(a) required 20 iterations by Method 1 and, as described above, the results of Fig. 3 required only 9 iterations by Method 1(a) followed by 3 more by Method 2(b).

At this writing, the program has not yet been written for the above formulation that includes the second-order-accurate formulation in Method 2. It is expected that when this is done, the program can be run rapidly with the first-order ($\lambda = 1$) operators on the very coarse mesh using Method 1, then switched to second-order ($\lambda = 2$) and Method 2 for final extrapolation.

5. Concluding Remarks

It has been shown that a special procedure (Method 2) is effective for obtaining most appropriate successive approximations for use in the Aitken extrapolation formula for accelerating the iterative convergence of numerical solutions to nonlinear partial differential equations. The procedure is based on the combined use of artificial perturbation-series expansions and an artificially extended equation. It was shown in a previous paper [8] that one version of the technique was very effective for accelerating iterative convergence when the solutions are smooth. The method, in a modified version, has now been applied with some success to a strongly supercritical transonic flow problem, in which the flow equations are of mixed type and whose solutions have shock-wave discontinuities. The method is expected to be extended to more general flows, including lifting airfoils and three-dimensional flows.

References

1 Aitken, A.C.: On Bernoulli's numerical solution of algebraic equations. Proc. Royal Soc. Edinburgh 46 (1926) 289–305.

2 Henrici, P.: Elements of Numerical Analysis. John Wiley & Sons, Inc., New York, 1964.

3 Shanks, D.: Nonlinear transformations of divergent and slowly convergent sequences. J. Math. and Phys. 34 (1955) 1–42.

4 Johnson, R. C.: Alternative approach to Padé approximants. Padé Approximants and Their Applications (ed. by P. R. Graves-Morris). Academic Press, London, 1973; 53–67.

5 Genz, A. C.: Applications of the ϵ-algorithm to quadrature problems. Padé Approximants and Their Applications (ed. P. R. Graves-Morris). Academic Press, London, 1973; 105–115.

6 Wynn, P.: On a device for computing the $e_m(S_n)$ transformation. Mathematical Tables and Other Aids to Computation 10 (1956) 91–96.

7 Wynn, P.: Acceleration techniques for iterated vector and matrix problems. Mathematics of Computation 16 (1962) 301–322.

8 Martin, E. D.; Lomax, H.: Rapid finite-difference computation of subsonic and slightly supercritical aerodynamic flows. Presented as a portion of AIAA Paper No. 74-11, 1974. Also AIAA Jour. 13 (1975) 579–586.

9 Martin, E. D.: Progress in application of direct elliptic solvers to transonic flow computations. Aerodynamic Analyses Requiring Advanced Computers, Part II, NASA SP-347, 1975; 839–870.

10 Martin, E. D.: A fast semidirect method for computing transonic aerodynamic flows. AIAA 2nd Computational Fluid Dynamics Conference Proceedings, 1975; 162–174.

11 Bellman, R.: Perturbation Techniques in Mathematics, Physics, and Engineering. Holt, Rinehart and Winston, New York, 1964.

12 Buneman, O.: A compact non-iterative Poisson solver. SUIPR Rept. 294, Inst. for Plasma Research, Stanford Univ., Stanford, Calif., 1969.

13 Hockney, R. W.: The potential calculation and some applications. Methods in Computational Physics, Vol. 9 (ed. by B. Alder, S. Fernbach, and M. Rotenburg). Academic Press, New York, 1970; 135–211.

14 Buzbee, B. L.; Golub, G. H.; Nielson, C. W.: On direct methods for solving Poisson's Equations. SIAM J. Numer. Anal. 7 (1970) 627–656.

15 Lomax, H.; Martin, E. D.: Fast direct numerical solution of the nonhomogeneous Cauchy-Riemann equations. J. Comp. Phys. 15 (1974) 55–80.

16 Hafez, M. M.; Cheng, H. K.: Convergence acceleration and shock fitting for transonic aerodynamics computations. AIAA Paper No. 75-51, 1975.

17 Concus, P.; Golub, G. H.: Use of fast direct methods for the efficient numerical solution of nonseparable elliptic equations. SIAM J. Numer. Anal. 10 (1973) 1103–1120.

18 Lomax, H.; Martin, E. D.: Variants and extensions of a fast direct numerical Cauchy-Riemann solver, with illustrative applications. NASA TN D-7934, 1975.

19 Murman, E. M.; Cole, J. D.: Calculation of plane transonic flows. AIAA Jour. 9 (1971) 114–121.

20 Murman, E. M.; Krupp, J. A.: Solution of the transonic potential equation using a mixed finite difference system. Lecture Notes in Physics 8 (ed. by M. Holt). Springer-Verlag, Berlin, 1971; 199–205.

21 Murman, E. M.: Analysis of embedded shock waves calculated by relaxation methods. AIAA Jour. 12 (1974) 626–633.

22 Murman, E. M.; Bailey, F. R.; Johnson, M. H.: TSFOIL-A computer code for 2-D transonic calculations, including wind-tunnel wall effects and wave-drag evaluation. Aerodynamic Analyses Requiring Advanced Computers, Part II, NASA SP-347, 1975; 769–788.

THE RISE OF A BUBBLE IN A FLUID

John L. Gammel
Department of Physics, Saint Louis University
St. Louis, Missouri 63103

I. Introduction

The standard numerical approach to problems in hydrodynamics is to replace the appropriate partial differential equations by a set of finite difference equations. This approach has several difficulties inherent in it. For example, fictitious viscosities are introduced and the cumulative effect of these viscosities may result in large errors for late times.

As a new approach, we advocate the exact solution of the partial differential equations by means of power series expansions.[1] This approach has the disadvantage that the power series may not converge for late times or in regions of space where the flow pattern varies rapidly. However, by use of Padé approximants or powerful generalizations of these approximations which we have developed for application to hydrodynamics, this difficulty has been overcome.

The problem which we consider as an illustration of our method is that of the rise of an incompressible volume of gas (that is, a bubble) which is initially spherical. The problem is to calculate the shape of the boundary and the height of the center of gravity at subsequent times. The problem has axial symmetry about the vertical or z-axis. To label a point on the boundary, we use the polar angle θ which is the angle between the vertical axis and a line joining the origin (the center of the initial sphere) and the point. The sphere has initially unit radius. At subsequent times, the shape of the boundary is specified by

$$\begin{aligned} x &= x(t, \cos\theta) \\ y &= y(t, \cos\theta) \end{aligned} \tag{1}$$

By known methods of hydrodynamics x and y are obtained as power series,

$$y = \cos\theta + t^2\left(-\frac{1}{4} + \frac{3}{4}\cos 2\theta\right) + t^4\left(-\frac{9}{32}\cos\theta + \frac{3}{32}\cos 3\theta\right)$$

$$+ t^6\left(\frac{9}{160} + \frac{1}{32}\cos 2\theta\right) + \dots ,$$

and x is best calculated from

$$\frac{\partial x}{\partial \theta}\frac{\partial^2 x}{\partial t^2} + \frac{\partial y}{\partial \theta}\left(\frac{\partial^2 y}{\partial t^2} + \frac{1}{2}\right) = 0 \quad ,$$

which expresses the fact that all forces other than gravity are perpendicular to the boundary.

Section II deals with methods of summing these series and the information which may be gained about the rise of the bubble and its shape at various times.

II. Methods of Series Summation

A. The ordinary Padé approximant.

Consider the function

$$f(x) = \sqrt{\frac{1+2x}{1+x}} = 1 + \frac{1}{2}x - \frac{5}{8}x^2 + \dots \; . \tag{3}$$

Suppose one has only the first three terms in this expansion and wants the value of $f(x)$ for some large positive value of x, say $x = \infty$. Substituting $x = \infty$ into the series will <u>not</u> result in a rapidly converging expression! [Were it known that $f(x) = \sqrt{(1+2x)/(1+x)}$, the source of the trouble would be obvious because $x = -1/2$ and $x = -1$ are branch points of $f(x)$ so that the series cannot converge for $|x| > 1/2$.] The Padé approximate method consists in writing $f(x)$ as the ratio of two polynomials,

$$f(x) = \frac{N_0 + N_1 x + \dots + N_m x^m}{1 + D_1 x + \dots + D_n x^n} \quad , \tag{4}$$

where $D_0 = 1$ by choice (were $D_0 \neq 1$, the numerator and denominator could be divided by it with the consequence $D_0 = 1$), and fixing the values of the N's and D's so that were the right hand side of Eq. (4) expanded in a power series in x, the result would agree with the right hand side of Eq. (3), that is, the original power series, through order m + n. The result is called the [m/n] Padé approximant to $f(x)$. For example, for the function of Eq. (3), the [1/1] Padé approximant is

$$\frac{N_0 + N_1 x}{1 + D_1 x} = 1 + \frac{1}{2}x - \frac{5}{8}x^2 + \dots \; , \tag{5}$$

and cross multiplying and equating powers of x yields

$$\text{coefficient of } x^0: \quad N_0 = 1 \quad ,$$

$$\text{coefficient of } x^1: \quad N_1 = \frac{1}{2} + D_1 \quad ,$$

$$\text{coefficient of } x^2: \quad 0 = -\frac{5}{8} + \frac{1}{2} D_1 \, , \tag{6}$$

from which

$$D_1 = 5/4 \, ,$$

$$N_1 = 7/4 \, ,$$

$$N_0 = 1 \quad . \tag{7}$$

Thus

$$\sqrt{\frac{1 + 2x}{1 + x}} \approx \frac{1 + \frac{7}{4} x}{1 + \frac{5}{4} x} \quad . \tag{8}$$

For $x = \infty$,

$$\sqrt{2} \approx 1.4 \, . \tag{9}$$

The source of this surprising accurate result is that $\sqrt{(1+2x)/(1+x)}$ has a cut, that is, a dense sequence of zeros and poles, running from $x = -1/2$ to $x = -1$. The Padé approximant has only one zero, at $x = -4/7$, and one pole, at $x = -4/5$, both of which lie between $x = -1/2$ and $x = -1$. The zeros and poles are rather like electric charges: when viewed from a great distance a complicated distribution of zeros and poles looks rather like a single zero and pole. The point $x = \infty$ is so far away from the cut that a single zero and pole suffice to represent the actual analytic structure very well.

Of course, one never relies only on the [1/1] Padé approximant: one discusses the convergence of a sequence of Padé approximants (in our example, the [1/1]. [2/2], [3/3], ... sequence). From Eq. (8) and also Eq. (4), one notes that these approximants are finite at $x = \infty$. We have used the information that $\sqrt{(1+2x)/(1+x)}$ is finite at $x = \infty$. In physical problems, such information is usually available. The [m/n] Padé approximants form a square array, and the approximants to be selected from this array is dictated by physical information.

B. The location of the zeros and poles.

In the example of Eq. (3), the zeros and poles of the [1/1], [2/2], [3/3], ... approximants lie between -1/2 and -1, and they become more dense as the order of approximation increases; that is, they form a cut. Cuts have no real meaning in analysis: in principle they can be located as one pleases so long as they join branch points. The Padé approximants have a definite opinion of their own about where the cuts go.

In many applications, the location chosen by the Padé approximant is at best awkward. Consider

$$f(x) = \sqrt{(1 - x)^2 + 1} \quad . \tag{10}$$

This function has branch points at $x = 1 \pm i$. The Padé approximants cut the function along an arc of circle running from $1 + i$ to^2 to $1 - i$ (the circle is entered at $x = 1$ and also passes through the origin). This situation is illustrated in figure 1. Beyond

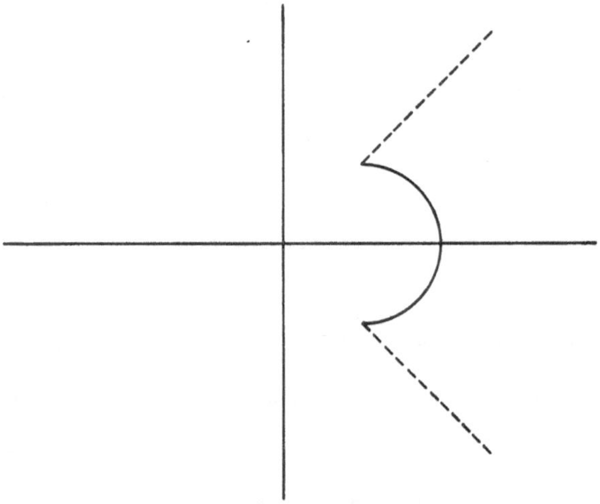

Figure 1

$x = 2$, the Padé approximants will not converge to $f(x)$, and even if they did, it could not be to that value of $f(x)$ on the principal Riemann sheet.

A way of moving these cuts is the following. This is very important for applications, because such a phenomenon is encountered in the summation of the expression for the height of the center of gravity of the bubble (see Eq. (2)).

This fact is reported at length in a Mission Research Corporation report. (An Accurate Early Time Solution for a Rising Fireball Model, by C. Longmire, G. McCartor, N. Carron, and F. Fajen, Report Number DNA 2967T, October, 1972 (MRC-R- 20).)

The scheme for moving the cut is very simple. Consider the even powers and odd powers separately, that is, consider

$$F_+(x^2) = \frac{1}{2}(f(x) + f(-x)) \quad ,$$
$$xF_-(x^2) = \frac{1}{2}(f(x) - f(-x)) \quad . \tag{11}$$

The location of the branch points is not changed, so that the singularities of F_+ and F_- are at $x^2 = (1\pm i)^2 = \pm 2i$. Now the zeros and poles of the Padé approximants to F_+ and F_- (whose numerators and denominators are now polynomials in x^2) are located on the imaginary axes, or, in the x-plane, along the dashed rays shown in Figure 1. Now the Padé approximants converge for all real positive x.

Actual numerical data will be presented in Section III.

C. Padé approximants for functions of two variables.

J. S. R. Chisholm of the University of Kent, Canterbury, has proposed the following extension of the defining equation (Eq. (4)) for functions of two variables:

$$f(x,y) = \sum_{\mu=0}^{N} \sum_{\nu=0}^{N} a_{\mu\nu} x^\mu y^\nu / \sum_{\sigma=0}^{N} \sum_{\tau=0}^{N} b_{\sigma\tau} x^\sigma y^\tau \ . \tag{12}$$

As before, we may choose $b_{00} = 1$. The a's and b's are defined by requiring that when the right hand side of Eq. (12) is expanded in a power series in x and y the result agrees with the power series expansion of f(x,y) through order 2N, and that in order 2N + 1, the average of the coefficients of $x^\gamma y^{2N+1-\gamma}$, $\gamma = 1,2,\ldots,N$, also agree.

This verbally complex algorithm is best explained with an example:

$$f(x,y) = c_{00} + c_{10}x + c_{01}y + c_{20}x^2 + c_{11}xy + c_{02}y^2$$

$$+ c_{30}x^3 + c_{21}x^2y + c_{12}xy^2 + c_{03}y^3 + \dots ,$$

$$\simeq \frac{a_{00} + a_{10}x + a_{01}y + a_{11}xy}{1 + b_{10}x + b_{01}y + b_{11}xy} . \tag{13}$$

Cross multiplying and equating the coefficients of various powers of x and y yields

$$\begin{aligned}
\text{coefficient of } x^0y^0&: \quad c_{00} = a_{00} ,\\
x^1y^0&: \quad c_{00}b_{10} + c_{10} = a_{10} ,\\
x^0y^1&: \quad c_{00}b_{01} + c_{01} = a_{01} ,\\
x^1y^1&: \quad c_{00}b_{11} + c_{10}b_{01} + c_{01}b_{10} + c_{11} = a_{11} ,\\
x^2y^0&: \quad c_{10}b_{10} + c_{20} = 0 ,\\
(x^1y^2 + x^2y^1)&: \quad c_{20}b_{01} + c_{11}b_{10} + c_{11}b_{01} + c_{02}b_{10} = 0 .
\end{aligned} \tag{14}$$

These linear equations for the a's and b's generally have one and only one solution.

The following properties of these approximants make them the most desirable of the possible approximants to functions of two variables:

1) The N/N Padé approximant to f(x)g(y) is the product of the ordinary N/N Padé approximants to f(x) and g(y) separately.

2) The N/N Padé approximant to f(x) + g(y) is the sum of the ordinary N/N Padé approximants to f(x) and g(y) separately.

D. Movement of cuts in functions of two variables.

Exactly the same tricks for moving cuts as that used in Section B for functions of one variable may be used for functions of two variables. In analogy with Eq. (11), one defines

$$F_{++}(x^2,y^2) = \frac{1}{4}\,(f(x,y) + f(-x,y) + f(x,-y) + f(-x,-y)) \quad ,$$

$$yF_{+-}(x^2,y^2) = \frac{1}{4}\,(f(x,y) + f(-x,y) - f(x,-y) - f(-x,-y)) \quad ,$$

$$xF_{-+}(x^2,y^2) = \frac{1}{4}\,(f(x,y) - f(-x,y) + f(x,-y) - f(-x,-y), \quad ,$$

$$xyF_{--}(x^2,t^2) = \frac{1}{4}\,(f(x,y) - f(-x,y) - f(x,-y) + f(-x,-y)) \quad . \tag{15}$$

It is important to note that the structure of the expansion of x and y (see Eq. (2)) suggests such a decomposition of these functions, because a general two variable expansion has 1 term of order 0, 2 terms of order 1, 3 terms of order 2, and so on, whereas these expansions of Eq. (2) have missing terms. Taking every other term, two expansions with the correct structure is found. When we come to the calculation of torus time in Section II, we will find that taking every other term results in Padé approximants whose poles and zeros are not awkwardly located.

E. Other generalizations of Padé Approximants.

One sees that Eq. (4) is nothing more or less than

$$Df - N = 0(x^{m+n}) \quad . \tag{16}$$

This is a linear equation in f. One might also consider

$$Pf^2 + Qf + R = 0(x^{m+n+\ell+1}) \quad , \tag{17}$$

where m, n, and ℓ are the degrees of polynomials P, Q, R, respectively. Such approximants are called quadratic Padé approximants. But, more important, we might consider

$$P\,\frac{d^2f}{dx^2} + Q\left(\frac{df}{dx}\right)^4 + R = 0(x^{m+n+\ell+1}) \quad , \tag{18}$$

because, if the power series expansion of f is known, so is the power series expansion of d^2f/dx^2, $(df/dx)^4$, etc., so that Eq. (18) results in linear equations for the coefficients in the polynomials P, Q, R.

III. The Bubble Problem.

A. Approximations based on partial differential equations.

From the series of Eq. (2), we obviously know x(t, cos θ) and y(t, cos θ), where x and y are the Cartesian coordinates of the point θ at time t (remember that θ refers to the initial configuration).

Had we great faith in Newtonian mechanics, we might very well start out a system of partial differential equations

$$\frac{\partial^2 x}{\partial t^2} + \ldots = 0 \ ,$$
$$\frac{\partial^2 y}{\partial t^2} + \ldots = 0 \ . \tag{19}$$

But what forces act on these particles? First, gravity

$$\frac{\partial^2 x}{\partial t^2} + \ldots = 0 \ ,$$
$$\frac{\partial^2 y}{\partial t^2} + \frac{1}{2} \ldots = 0 \ , \tag{20}$$

where y is the vertical axis. What else? The pressure acts perpendicularly to the boundary, and since the magnitude of the pressure ought to be proportional to y, and the direction cosines of a vector perpendicular to the boundary are $+\partial y/\partial\theta$, $-\partial x/\partial\theta$,

$$\frac{\partial^2 x}{\partial t^2} + \quad + \alpha y \frac{\partial y}{\partial \theta} + \ldots = 0 \ ,$$
$$\frac{\partial^2 y}{\partial t^2} + \frac{1}{2} \quad - \alpha y \frac{\partial x}{\partial \theta} + \ldots = 0 \ . \tag{21}$$

There are properties that such a set of equations ought to have. It ought to be

1) Galilean invariant, that is, invariant against $y = y' + vt$, and
2) invariant against transformations of any sort on the variable θ,

$$\theta = \theta(\theta') \ . \tag{22}$$

The pressure term is not Galilean invariant, but it can be made invariant against transformations of θ by writing instead

$$\frac{\partial^2 x}{\partial t^2} + \quad + \alpha y \frac{\partial y/\partial\theta}{\sqrt{\left(\frac{\partial x}{\partial\theta}\right)^2 + \left(\frac{\partial y}{\partial\theta}\right)^2}} + \ldots = 0 \quad ,$$

$$\frac{\partial^2 y}{\partial t^2} + \frac{1}{2} - \alpha y \frac{\partial x/\partial\theta}{\sqrt{\left(\frac{\partial x}{\partial\theta}\right)^2 + \left(\frac{\partial y}{\partial\theta}\right)^2}} + \ldots = 0 \quad . \tag{23}$$

To make these Galilean invariant, we need to impose the incompressibility as a constraint. According to d'Alembert's principle,

$$\left(\frac{\partial^2 x}{\partial t^2} + \ldots\right) dx + \left(\frac{\partial^2 y}{\partial t^2} + \ldots\right) dy = 0 \quad , \tag{24}$$

and a constraint V(x,y) = constant has to be handled by writing

$$\lambda\frac{\partial V}{\partial x}\, dx + \lambda\frac{\partial V}{\partial y}\, dy \quad , \tag{25}$$

where λ is a Lagrange multiplier, and subtracting Figs. (24) and (25), so that

$$\frac{\partial^2 x}{\partial t^2} + \lambda\frac{\partial V}{\partial x} + \ldots = 0 \quad ,$$

$$\frac{\partial^2 y}{\partial t^2} + \lambda\frac{\partial V}{\partial y} + \ldots = 0 \quad . \tag{26}$$

A consideration too long to be included here shows that the constraint (volume = constant) results in

$$\frac{\partial^2 x}{\partial t^2} \quad + \lambda \frac{\frac{\partial y}{\partial\theta}}{\sqrt{\left(\frac{\partial x}{\partial\theta}\right)^2 + \left(\frac{\partial y}{\partial\theta}\right)^2}} + \alpha y \frac{\frac{\partial y}{\partial\theta}}{\sqrt{\left(\frac{\partial x}{\partial\theta}\right)^2 + \left(\frac{\partial y}{\partial\theta}\right)^2}} + \ldots = 0 \quad ,$$

$$\frac{\partial^2 y}{\partial t^2} + \frac{1}{2} - \lambda \frac{\frac{\partial x}{\partial\theta}}{\sqrt{\left(\frac{\partial x}{\partial\theta}\right)^2 + \left(\frac{\partial y}{\partial\theta}\right)^2}} - \alpha y \frac{\frac{\partial y}{\partial\theta}}{\sqrt{\left(\frac{\partial x}{\partial\theta}\right)^2 + \left(\frac{\partial y}{\partial\theta}\right)^2}} + \ldots = 0 \quad . \tag{27}$$

It is very important that λ is a function of t only. The Galilean invariance is now obvious because when $y = y' + vt$, the vt term is absorbed in the Lagrange multiplier, so that $\lambda' = \lambda + \alpha vt$.

What other types of terms might one consider adding? "Drag" terms proportional to $v_\perp{}^2$, where $v_\perp$ is the component of the velocity perpendicular to the boundary, or $v_{11}{}^2$, and perpendicular to the boundary, might be added. Such terms are Galilean invariant because we interpret $v_\perp$ to be measured relative to the fluid at rest at infinity.

Our final equations are

$$\frac{\partial^2 x}{\partial t^2} + \lambda\frac{\frac{\partial y}{\partial\theta}}{D} + \alpha y\,\frac{\frac{\partial y}{\partial\theta}}{D} + \beta v_\perp{}^2\,\frac{\frac{\partial y}{\partial\theta}}{D} + \gamma v_{11}{}^2\,\frac{\frac{\partial y}{\partial\theta}}{D} = 0 \ ,$$

$$\frac{\partial^2 y}{\partial t^2} + \frac{1}{2} - \lambda\frac{\frac{\partial x}{\partial\theta}}{D} - \alpha y\,\frac{\frac{\partial x}{\partial\theta}}{D} - \beta v_\perp{}^2\,\frac{\frac{\partial x}{\partial\theta}}{D} - \gamma v_{11}{}^2\,\frac{\frac{\partial x}{\partial\theta}}{D} = 0 \ , \qquad (28)$$

$$D = \sqrt{\left(\frac{\partial x}{\partial\theta}\right)^2 + \left(\frac{\partial y}{\partial\theta}\right)^2} \ .$$

We determine α, β, and γ so that the expansion of the solutions of Eq. (28) in a double power series in t and cos θ agrees with the correct expansions (see Eq. (2)) to as high an order as possible. With

$$\alpha = \frac{3}{2} \ , \qquad \beta = -\frac{33}{16} \ , \qquad \gamma = -\frac{15}{4} \ , \qquad (29)$$

agreement is attained through order t^6 (all powers of cos θ relevant to order t^6). The choice of the exponents in $v_\perp{}^2$ and $v_{11}{}^2$ are also dictated by requiring this agreement, and the power series expansion of λ(t) is also fixed.

Some figures for the rise of the bubble as calculated in this manner are shown in Appendix I (figures 1-6). Torus time comes quite early, at t = 1.095 or so.

It is extremely difficult to generalize this idea to higher powers of t, and we have not done that. But by abandoning the invariance requirements and going back to

$$A\,\frac{\partial^2 x}{\partial t^2} + \frac{3}{2}\,y\,\frac{\frac{\partial y}{\partial\theta}}{D} + \lambda\,\frac{\frac{\partial y}{\partial\theta}}{D} - B\,\frac{33}{16}\,v_\perp{}^2\,\frac{\frac{\partial y}{\partial\theta}}{D} - C\,\frac{\frac{\partial y}{\partial\theta}}{D} = 0 \ ,$$

$$A\,\frac{\partial^2 y}{\partial t^2} + \frac{1}{2} - \frac{3}{2}\,y\,\frac{\frac{\partial x}{\partial\theta}}{D} - \lambda\,\frac{\frac{\partial x}{\partial\theta}}{D} + B\,\frac{33}{16}\,v_\perp{}^2\,\frac{\frac{\partial x}{\partial\theta}}{D} + C\,\frac{\frac{\partial x}{\partial\theta}}{D} = 0 \ , \qquad (30)$$

we can fix the coefficients in the polynomials A, B, and C so that agreement through any order in t is attained. The danger is that the polynomial A will develop zeros, but in some orders and at some places on the boundary this danger is not realized. Torus time tends to come later as the degree of approximation (that is, as the order of the polynomials A, B, C) is increased, perhaps at t = 1.2 or so, as figure 7 exhibited in Appendix I shows.

B. Estimate of torus time by means of ordinary Padé approximants.

The thickness 2ℓ of the bubble is, of course, the difference in y at the top of the bubble and y at the bottom, that is

$$2\ell = y(t^2, 0) - y(t^2, \pi) \quad . \tag{31}$$

Explicitly,

$$\ell = 1 - 1.875 \times 10^{-1}\, t^4 - 5.438 \times 10^{-2}\, t^8 - 1.362 \times 10^{-2}\, t^{12} + \ldots \quad . \tag{32}$$

These coefficients are tabulated to a large number of figures in Table I. The [1/1], [2/2], [3/3],... Padé approximants yield $\ell = 0.767$ for t = 1 in good agreement with the results obtained from the partial differential equation approach. In general, they all give $\ell = 0$ (Torus time) at t = 1.2 or so, also in good agreement with the partial differential equation approach (see Table II).

Now, it could be that the zero in ℓ is a zero located on a cut crossing the real t axis (see the location of poles and zeros in Table II). To study this possibility, we move the cut as discussed in section II.B, that is we write

$$\ell = F_+ (t^8) + t^4\, F_- (t^8) \quad . \tag{33}$$

Various Padé approximants to F_+ and F_- at various times are shown in Table III. In general, they suggest $\ell = 0$ at some time later than t = 1.2, perhaps t = 1.4 or even 1.5. They also appear to have no undesirable zeros or poles, and the approximations seem to change quite smoothly as the order of approximation increases.

Really, although it may appear in section II that an enormous number of terms have been calculated in the expansions for $y(t^2, \theta = 0)$ or $y(t^2, \theta = \pi)$, we are reduced in Eq. (33) to functions whose expansion coefficients are every fourth term in the original expansion, and 36/4 = 9 so that only a [4/5] or [5/4] Padé approximant may be studied. More work is needed to generate accurate series expansions to higher order; such work will require much computer time on fast computers.

The partial differential equation method is applicable to the rise of a cylindrical bubble, which develops a cap after torus time; that is, it looks like

at torus time. According to Gary McCartor, the series expansions needed for the cylindrical bubble are

$$r = 1 + t^4\left(\frac{1}{6} - \frac{7}{18}\cos 2\phi\right) + t^6\left(-\frac{28}{135}\cos\phi + \frac{73}{108}\cos 3\phi\right)$$

$$\theta = \phi + t^2(\sin\phi) + t^4\left(-\frac{1}{18}\sin 2\phi\right) + t^6\left(-\frac{101}{1080}\sin\phi + \frac{35}{72}\sin 3\phi\right)$$

$$h = \frac{1}{2}t^2 + \frac{7}{90}t^6 + \ldots, \quad x = r\sin\phi, \quad y = r\cos\phi + h. \tag{34}$$

Acknowledgments

A great deal of the work reported here was done by Gary McCartor and William Wortman of the Mission Research Corporation, Santa Barbara, California, and Mary Menzel of the Los Alamos Scientific Laboratory.

References

1. For previous theoretical and experimental work see J. K. Walters and J. F. Davidson, J. Fluid Mech. 12, 408 (1962) and 17, 321 (1963).

Figure Captions

Figures 1-6. Shape of the bubble at various times. The time and distance scales are set by the initial rise during which the bubble behaves as though it were rigid rising like $\frac{1}{2}t^2$ and by the initial radius $r = 1$, respectively. $h(t)$ is the height of the point $x = 0$, $y = \frac{1}{2}y(0,t) + \frac{1}{2}y(\pi,t)$.

Figure 7. Thickness of the bubble on axis of symmetry vs time. These are best estimates from calculations based on Eq. (30).

APPENDIX I

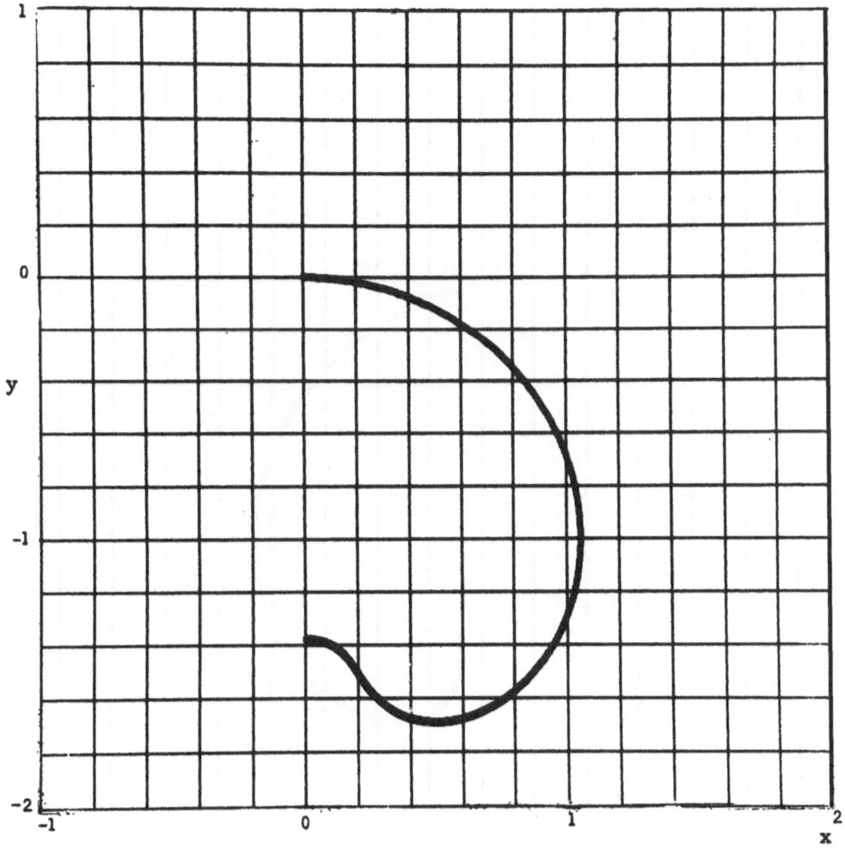

Figure 1

t = 1

h(t) = 1.372

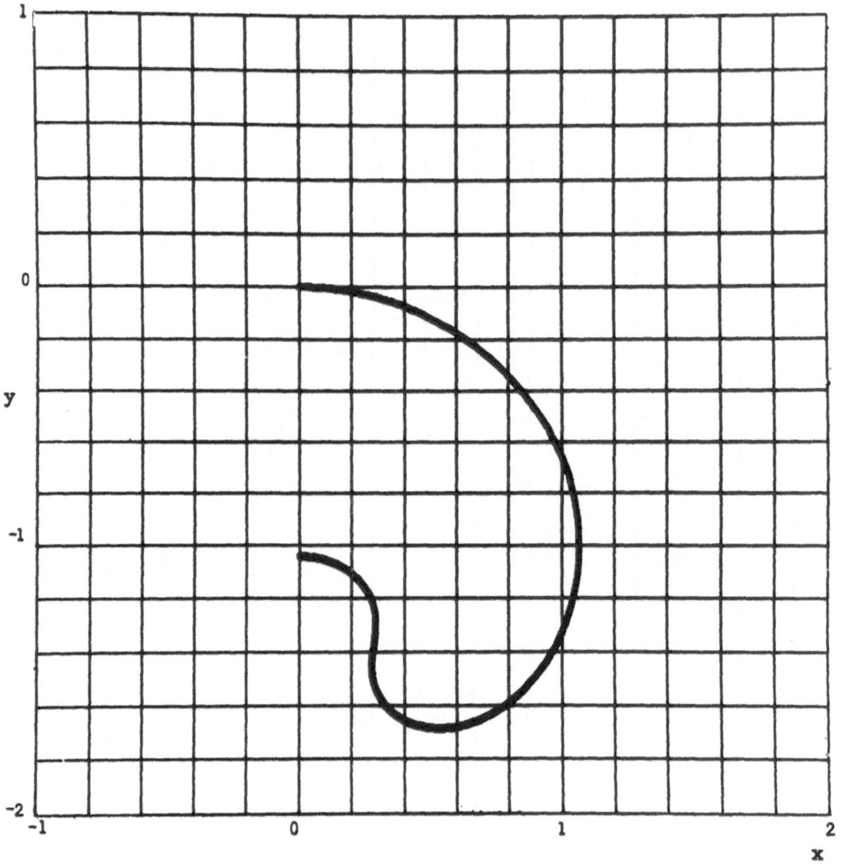

Figure 2

t = 1.05

h(t) = 1.406

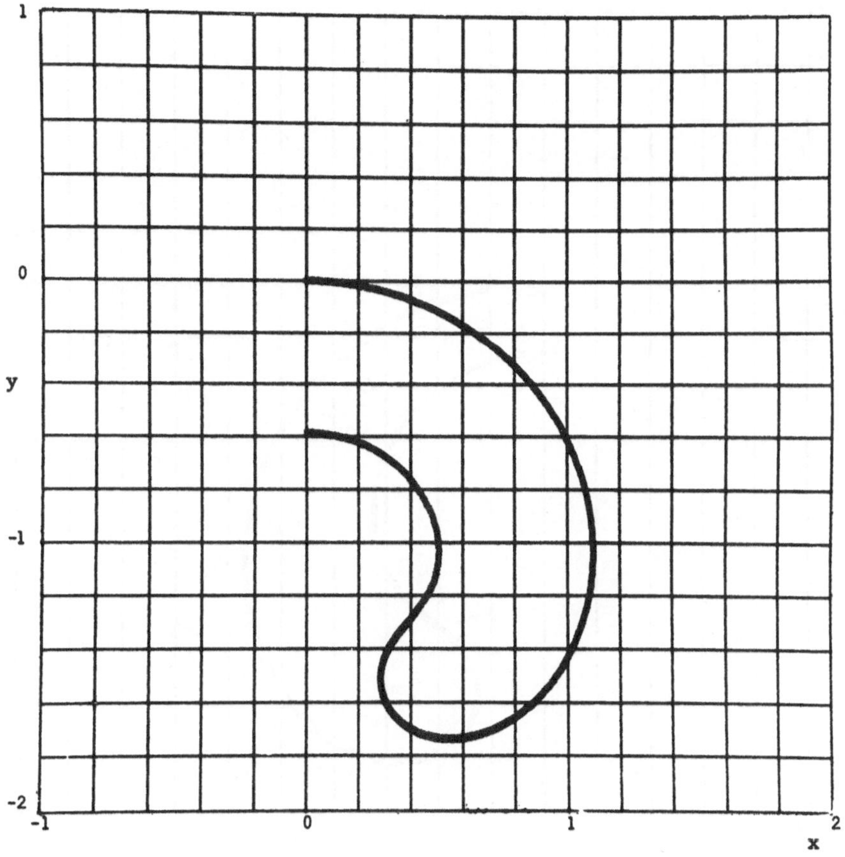

Figure 3

t = 1.08

h(t) = 1.450

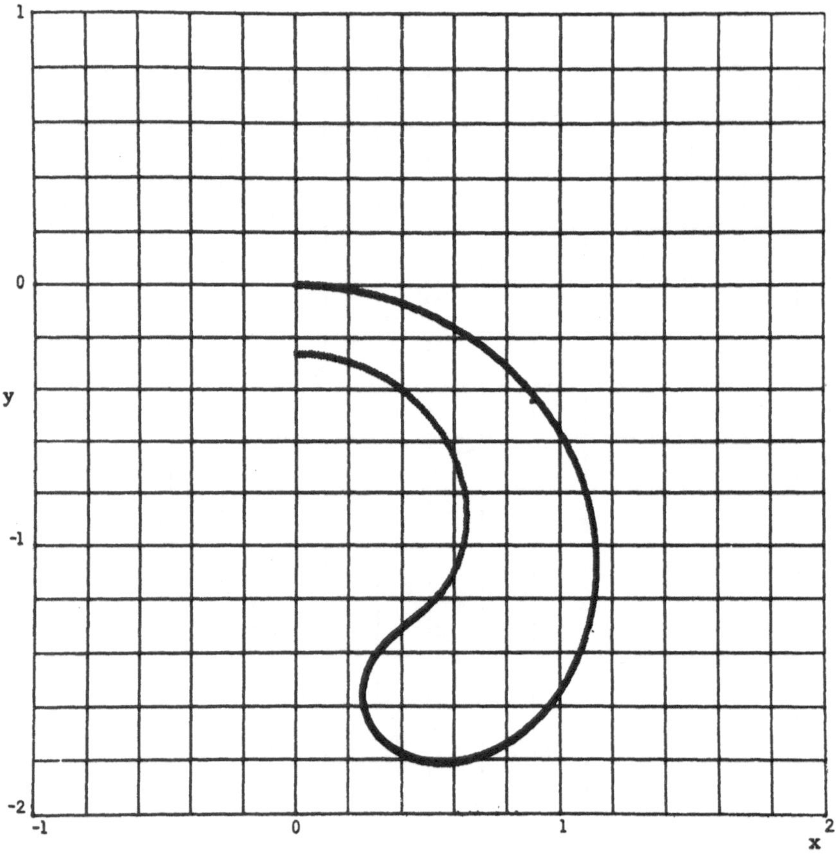

Figure 4

t = 1.09

h(t) = 1.493

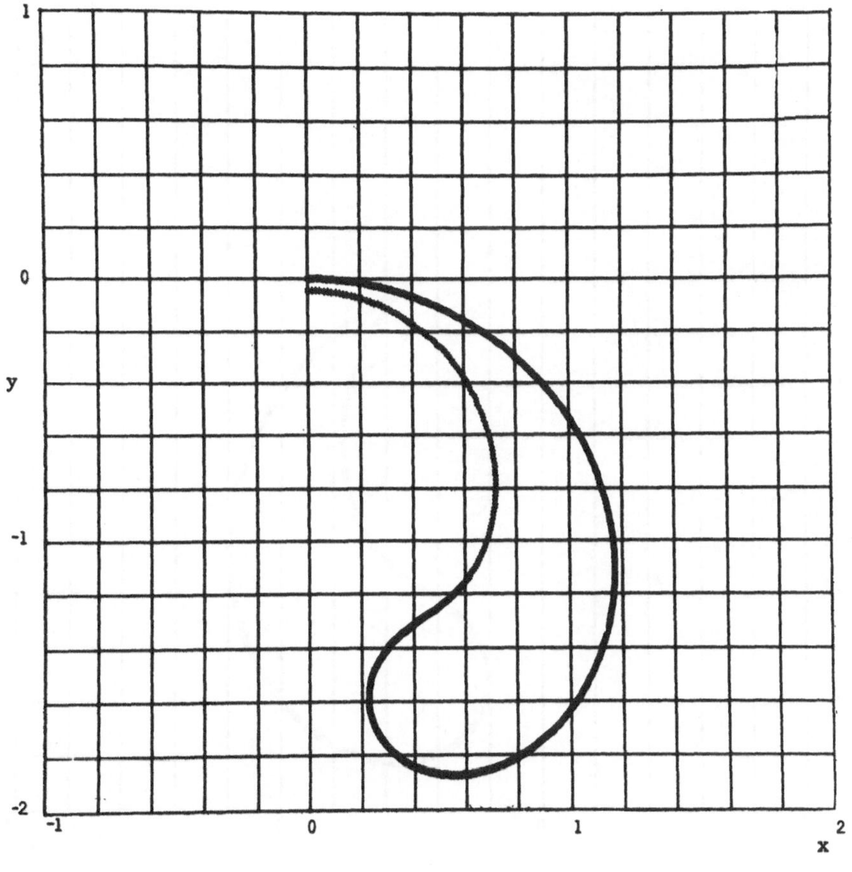

Figure 5

t = 1.094

h(t) = 1.527

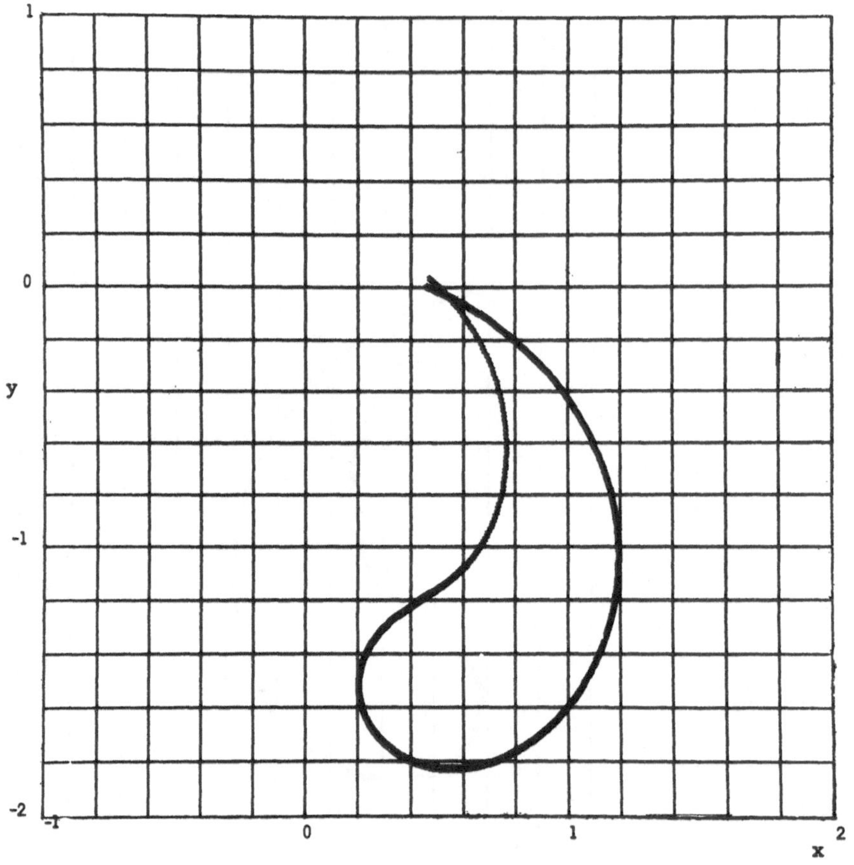

Figure 6

t = 1.095

h(t) = 1.530

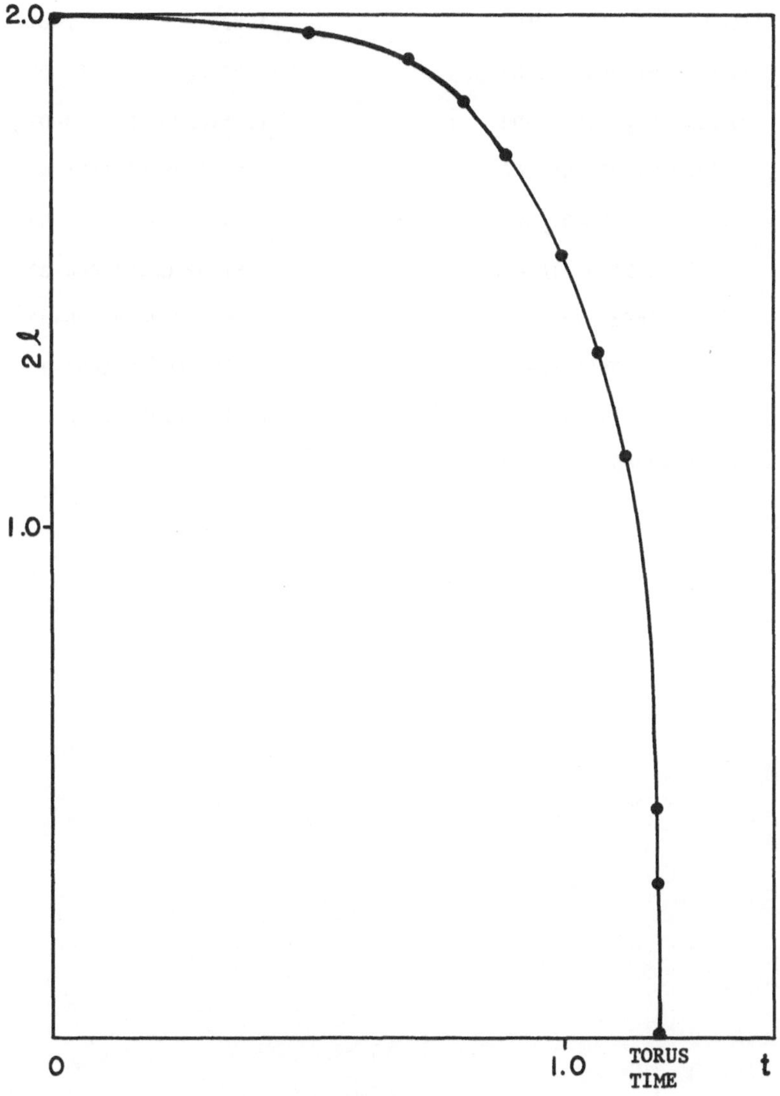

Figure 7

Thickness of the bubble on axis of symmetry vs time. These are best estimates from calculations based on Eq. (30).

Table I

Accurate Coefficients for Eq. (30). The order in which these are to be read is apparent from Eq. (30).

1.0 +00	-1.875 -01
-5.438456632653055-02	-1.362446243532089-02
5.11511872461845 -03	1.151623084926745-02
1.012249732304565-02	4.804358858144425-03
-9.952368481864343-04	-4.791099632885595-03
-5.506523714786576-03	-3.513584047019944-03
-2.120294683714934-04	2.692173228391798-03
3.926479535644956-03	3.128540287264697-03
8.953856051941572-04	-1.579142125257331-03
-3.09449704894324 -03	

Table II

Padé Approximants formed from Series of Eq. (30).
Half the thickness of the bubble on the axis of
symmetry is tabulated.

t	3/4	4/5	7/8	8/9
0.8	0.91347018	0.91347002	0.91347014	0.91347014
1.0	0.76469807	0.76420450	0.76424654	0.76685838
1.1	0.61689611	0.61116700	0.59803286	0.66098736
1.2	0.22093630	0.18768951	< 0	0.37334144
1.3	< 0	< 0		< 0

Location (in t^4 plane) of the zeros and poles
of the 8/9 Padé Approximant

Roots of Denominator	Roots of Numerator
0.75606093 ± i 0.64871509	0.75563195 ± i 0.64727296
0.92782164 ± i 0.71306521	0.92546957 ± i 0.70349188
1.3703819 ± i 0.78104964	1.3489190 ± i 0.73761275
2.9883484	2.3641068
14.699468	9.7374222
-139.05115	

Table III

Padé Approximants to F_+ (see Eq. (33))

t	0/1	1/2	2/3	3/4	4/5
0.8	0.99095828	0.99128592	0.99106605	0.99106605	0.99106605
1.0	0.94842056	0.94792005	0.95637556	0.95636353	0.95665642
1.1	0.89559338	0.89475280	0.92423556	0.92412823	0.92994608
1.2	0.81047544	0.80916506	0.86835050	0.86804717	0.89101745
1.3	0.69269749	0.69092524	0.77897013	0.77845245	0.82240395
1.4	0.55475379	0.55272305	0.6602585	0.65956997	0.72114121
1.5	0.41774289	0.41575710	0.52715917	0.52639470	0.59747092

t	1/0	2/1	3/2	4/3	5/4
0.8	0.99087578	0.99109132	0.99106605	0.99106609	0.99106605
1.0	0.94561543	0.94039026	0.95619283	0.95616968	0.95664655
1.1	0.88342185	0.87617212	0.92087900	0.92069538	0.92905483
1.2	0.76615632	0.75356214	0.84900828	0.84847787	0.88127337
1.3	0.55636838	0.53389113	0.71094835	0.70988429	0.78092139
1.4	0.19739852	0.15790015	0.46976099	0.46784701	0.59723664
1.5	-0.39381695	-0.46139448	0.07036295	0.06707928	0.28885363

Roots of Numerator (4/5)	Roots of Denominator (4/5)
$0.1077 \pm i\ 1.053$	$0.1053 \pm i\ 1.054$
$0.1410 \pm i\ 1.975$	$0.0982 \pm i\ 1.991$
	-36.932

Roots of Numerator (5/4)	Roots of Denominator (5/4)
$0.1205 \pm i\ 1.048$	$0.1183 \pm i\ 1.050$
$0.2617 \pm i\ 1.938$	$0.2240 \pm i\ 1.962$
35.761	

Table III (continued)

Padé Approximants to F_- (see Eq. (30))

t	1/1	2/2	3/3	4/4
0.8	-0.18950191	-0.18944294	-0.18944316	-0.18944316
1.0	-0.19488349	-0.18943211	-0.19004561	-0.18978128
1.1	-0.19788633	-0.18046367	-0.18498208	-0.18047145
1.2	-0.20014065	-0.17087111	-0.18157475	-0.16572802
1.3	-0.20157703	-0.16549046	-0.18068947	-0.15482310
1.4	-0.20242239	-0.16291690	-0.18070385	-0.14917041
1.5	-0.20290741	-0.16165605	-0.18088619	-0.14641120

Roots of Numerator (4/4)	Roots of Denominator (4/4)
$0.09447 \pm i\ 1.125$	$-0.09945 \pm i\ 2.511$
$-0.3624 \pm i\ 2.801$	$0.09920 \pm i\ 1.104$

RATIONAL APPROXIMATIONS TO THE SOLUTION OF THE BLUNT-BODY & RELATED PROBLEMS

L. W. SCHWARTZ
DEPARTMENT OF APPLIED MATHEMATICS
UNIVERSITY OF ADELAIDE, ADELAIDE, S.A.

I Introduction

A body travelling at a speed greater than the speed of sound in a compressible fluid will, in general, be preceded by a shock-wave. This "bow-shock" wave separates the fluid region into two parts: the fluid ahead of the oncoming body is unaware of the body's presence, while the fluid between the shock and the body is affected quite strongly. The problem of finding the shape of the bow-shock, the details of the fluid motion and the resulting pressures and forces on the body is usually referred to as the supersonic blunt-body problem. Interest in the problem has been strong for the past thirty years; currently the major application is to the design of spacecraft.

While the most general problem involves viscous, unsteady, chemical non-equilibrium, heat and mass transfer and other considerations, experience has shown that the solution of the "classical" blunt-body problem, involving the steady flow of a thermally and calorically perfect compressible fluid with viscous effects neglected, is adequate for many purposes.

The results presented in this paper employ an inverse method of solution to this problem. In an inverse method, the shape of the detached bow-wave is prescribed; the values of the flow variables immediately downstream of the bow shock can be specified in terms of the free-stream parameters by means of the Rankine-Hugoniot (shock jump) relations. These values serve as the initial conditions for the system of nonlinear partial differential equations that characterize the flow. The shape of the body which produces the given shock-wave is determined as part of the solution. The solution of the so-called direct problem, where the body shape and free-stream conditions are specified initially, requires, in general, an iterative computation. Because exact initial conditions can be specified only at the shock-wave, rather than at the body surface, all solutions to the blunt-body problem are, of necessity, indirect.

Numerical solutions to the inverse blunt-body problem have met with a good measure of success, especially for axisymmetric flow problems. These solutions usually involve a finite-difference, marching-from-the-shock technique. However, many instabilities plague these numerical solutions, caused both by the particular numerical procedure employed as well as by the mathematical nature of the inverse problem. An important example in the latter category arises because the flow behind any detached shock must be at least partially subsonic. In the subsonic region the governing equations are elliptic in character. An elliptic system of equations with initial (Cauchy) data is known to be mathematically ill-posed.[1] Slight errors in the initial data can lead to unacceptably large deviations in the solution.

An alternate, semi-analytic, solution to the blunt-body problem can be obtained via a series expansion method. Here the solution for any flow variable is assumed to be a (multiple) Taylor series in the space coordinates with undetermined coefficients. If the shape of the bow shock is given by an analytic relation, values of the flow variables and arbitrarily many derivatives of these variables running along the shock may be found at once from the Rankine-Hugoniot relations. The system of differential equations governing the motion may then be manipulated to yield arbitrarily many derivatives of the variables normal to the shock in terms of the derivatives along the shock.

It is interesting to note that the first attempts to solve the blunt-body problem, in the late forties and early fifties, employed this series approach[2,3,4]. Their results were inconclusive, however, partly because they computed their series by hand and thus were able to extract only five or six terms. Moreover, Van Dyke[5] discovered, in 1958, that a limit line appears in the upstream analytic continuation of the flow, that is, "ahead" of the shock. This limit line, while only a "mathematical fiction", will very frequently lie closer to the shock than does the body and will therefore determine the "radius" of convergence of the series expansion. A successful solution based on the series expansion method must therefore incorporate an analytic continuation procedure. Van Tuyl[6] suggested the use of Padé fractions for this purpose. This idea was used by Moran[7] along with automated computation of the series coefficients to produce successful blunt-body solutions for problems with axial symmetry.

In this paper we report on an extension of Moran's work to non-symmetric configurations. We also discuss two related problems which, because of their relative simplicity, allow one to elucidate certain features of the analytic structure of their solutions. We make use of a battery of techniques, including Domb-Sykes plots[8], Euler transformations and "series completion"; collectively these procedures have come to be known as "the method of Van Dyke" in acknowledgement of his pioneering work in this field.[9,10].

II The Taylor-Maccoll Conical Flow Solution

The problem of axially symmetric supersonic flow past a right circular cone was solved numerically by Taylor and Maccoll in 1933[11]. More refined numerical solutions have been produced by Kopal[12], Simms[13,14], and others. Provided the free-stream Mach number is high enough to permit an attached shock which is a coaxial circular cone, the flow quantities will depend only on one space coordinate, the azimuthal angle θ. Though the analogy is by no means complete, the Taylor-Maccoll problem may be considered to be a model "one-dimensional blunt-body problem" and, as such, merits study before proceeding to more general cases.

The problem is formulated in spherical polar coordinates with the axis of the cone corresponding to the origin of the azimuthal angle θ. Let w,u denote the velocity components in the r and θ directions, respectively. They are determined as the solution of the 2nd order nonlinear equation:

$$(a^2-u^2)\left(\frac{d^2w}{d\theta^2}+w\right)+a^2\left(w+\frac{dw}{d\theta}\cot\theta\right)=0 \qquad (1)$$

where

$$u=\frac{dw}{d\theta}$$

and

$$a^2+\frac{\gamma-1}{2}(u^2+w^2)=\text{const}=\frac{\gamma-1}{2}V_m^{\ 2} \qquad (2)$$

follows from the conservation of total enthalpy. Here $a=a(\theta)$ is the local speed of sound, γ the ratio of specific heats, and V_m the velocity the fluid would attain if allowed to expand adiabatically into a vacuum. The solid cone surface is characterized by zero normal velocity:

$$u(\theta_c)=0.$$

Thus if $w_c=w(\theta_c)$ is specified, the equations may be integrated numerically from the solid cone until a value of θ is attained where the velocity components are compatible with the Rankine-Hugoniot relations. Thus this problem, like the more general blunt-body problem, requires the satisfaction of mixed boundary conditions. The solution of the direct problem, i.e. free-stream Mach number and solid cone angle initially specified, will require an iterative computation.

In a later paper Maccoll[12] sought a power series solution for the flow. He assumed, in effect, a solution of the form

$$w=\sum_{j=0}^{\infty} w_j\varepsilon^j \qquad (3)$$

where $\varepsilon=\theta-\theta_c$ and substituted this expansion in (1). Choosing as his dimensionless dependent variable the quantity w/V_m, he obtained the series coefficients through $O(\varepsilon^5)$. These coefficients are in general functions of the three parameters θ_c, γ, and w_c/V_m. In Maccoll's formulation the parameter γ first appears in the $O(\varepsilon^4)$ term, as a factor $1/(\gamma-1)$, thus indicating a very strong dependence on the ratio of specific heats, particularly for polyatomic gases.

In a recent paper, this author[16] has shown, via a small modification to Maccoll's procedure, that the apparent strong dependence on γ is spurious. Choosing the cone velocity w_c as the reference, one obtains

$$\frac{w}{w_c}=1-\varepsilon^2+\frac{\varepsilon^3}{3}\cot\theta_c-\left(\frac{M_c^{\ 2}}{3}+\frac{1}{4}\cot^2\theta_c\right)\varepsilon^4$$

$$+\frac{\cot\theta_c}{20}\left(\frac{7}{3}+\frac{40}{3}M_c^2+4\cot^2\theta_c\right)\varepsilon^5$$

$$-\frac{\varepsilon^6}{30}\left[\frac{2}{3}+\frac{19}{4}\cot^2\theta_c+5\cot^4\theta_c\right.$$

$$\left.+M_c^2\left(\frac{14}{3}+\frac{89}{3}\cot^2\theta_c\right)+M_c^4\left(\frac{64}{3}+4(\gamma-1)\right)\right]+\ldots \qquad (4)$$

through order ε^6. Here it is seen that γ appears only in the ε^6 term and moreover that the effect on this coefficient of large changes in γ (from a physical viewpoint) will be quite minor. This suggests that the ratio w/w_c may be sensibly independent of γ and hence the nature of the gas. Exact solutions of (1) appear to bear out this conjecture. Indeed the greatest change in the normalized velocity ratio at any point in the flow when γ is increased from 1.4 to 5/3 (i.e. diatomic versus monatomic gas) is about one part in 3000. It seems fair to say that a type of "quasi-similarity" has been demonstrated and that the only sensible dependence on γ derives from the shock boundary condition.

The exact solutions used for these comparisons were obtained from high-order series solutions recast as rational fractions. Machine computation using the series approach is at least as efficient as standard finite difference methods for this problem. These series solutions can be made to yield information concerning the nature and location of important singularities through the use of Domb-Sykes plots. This graphical extension of the familiar D'Alembert ratio test is based on the following observation: If

$$f(\varepsilon)=\sum_{n=0}^{\infty}a_n\varepsilon_n=\begin{cases}k(\varepsilon_0\pm\varepsilon)^{\alpha} & \alpha\neq 0,1,\ldots\\ k(\varepsilon_0\pm\varepsilon)^{\alpha}\log\ (\varepsilon_0\pm\varepsilon) & \alpha=0,1,\ldots\end{cases} \qquad (5a)$$

then

$$\frac{a_n}{a_{n-1}}=\mp\frac{1}{\varepsilon_0}\left[1-\frac{1+\alpha}{n}\right] \qquad (5b)$$

and hence that the ratios of series coefficients, when plotted versus 1/n, will lie on a straight line. Thus if we make such a plot using the coefficients in any series and if the points ultimately tend towards a straight-line asymptote, we can infer that the closest singularity is of type (5a). The plot yields estimates of the nature α and location ε_0 of this singularity. We show such a plot formed from the coefficients in (3)in Fig.1. A singularity corresponding to a 3/2-power branch point at $\varepsilon_0 \simeq -.156$ is indicated. Here the singularity closest to the surface of the solid cone lies "buried within" the cone to a "depth" of about 8.9 degrees. In this case the shock lies closer to the cone surface than does the branch-point. Hence for $\theta_c = 40^0$ and $M_c = 2$, the shock will be within the series expansion from the solid cone. Very often this is not true, as for a slender cone at relatively low free-stream Mach number. The oscillations in the plot of Figure 1 are caused primarily by another singularity lying somewhere ahead of the shock. Information about this secondary singularity can be obtained after first mapping away the one at ε_0 by means of the Euler transformation

$$\hat{\varepsilon} = \frac{\varepsilon}{\varepsilon-\varepsilon_0} \quad . \tag{6}$$

A Domb-Sykes plot of the series for w ($\hat{\varepsilon}$) reveals that the outer discontinuity is also a 3/2-power branch-point. This appears to be true in general; the two singular lines correspond to the envelopes of right-and left-running characteristics where these both exist.

Using the information that we have obtained about the nature of the important singularities of w and the virtual lack of dependence on γ, it is instructive to examine the accuracy of a low-order Padé approximant formed from the first few series coefficients. Since we know that the important singularities of the Taylor-Maccoll solution are 3/2-power branch-points, a simple modification can result in an improved approximation. The series solution (4) is raised to the 2/3 power; the new power series is then cast as a rational fraction; finally, the rational fraction is raised to the 3/2 power to recover an approximation to w. Thus each zero of the rational fraction is converted to a branch-point of the proper kind. Using only four terms of (4), we form

$$\begin{aligned} w(\varepsilon;M_c,\theta_c) &\cong \left\{ [3/1]\ (w^{2/3}) \right\}^{3/2} \\ &= \left\{ 1 - \frac{2}{3}\varepsilon^2 \left[\frac{1+[(1-6w_4)/6w_3-w_3]\ \varepsilon}{1+[(1-6w_4)/6w_3]\ \varepsilon} \right] \right\}^{3/2} \end{aligned} \tag{7}$$

where the mnemonic notation [3/1] signifies a rational fraction with three zeroes and one pole. w_3 and w_4 are coefficients from (4). Figure 2 compares the approximate normalized profiles obtained from (7) with exact results for a cone of 5^0 half-angle at both a high and a low surface Mach number. The greatest error is about 4% for the low Mach number case. For $M_c = 10$, on the other hand, the maximum error is less than 1%. Two other case comparisons, for $\theta_c = 40^0$, likewise show agreement to within 1 percent. (See reference 16).

The relatively large error for the small Mach-number and cone-angle case can be explained, and, in principle, removed by a method of "partial series completion". In (4) we observe that each coefficient w_j contains a term

$$(-1)^{j+1}\ \frac{\cot^{j-2}\theta_c}{j} \tag{8}$$

for j=2,3... From the Taylor series expansion

$$\frac{\log(1+\varepsilon\cot\theta_c)}{\cot^2\theta_c} - \frac{\varepsilon}{\cot\theta_c} = \sum_{j=2}^{\infty} \frac{(-1)^{j+1}\cot^{j-2}\theta_c}{j}\ \varepsilon^j \tag{9}$$

one may infer the presence of a logarithmic singularity at $\varepsilon = -\tan\theta_c$.
Assuming that all the coefficients in the series contain terms of the form (8), the

series may be partially completed and the logarithmic singularity removed through the use of (9). For θ_c<<1 the logarithmic singularity is only slightly further from the surface than the 3/2-power branch point. Because it is stronger than the branch point, however, it dominates the leading portion of the series (4).

Thus by using various bits of information extracted from the leading terms in a series solution to the Taylor-Maccoll problem, we are able to case light on the analytic structure of the full solution and obtain a simple and accurate approximation. It should be emphasised that (7) is a rational approximation derived from analytic principles, rather than a curve fit and should therefore by systematically improved when carried to higher order.

III Hypersonic Flows with Parabolic Shocks

Perhaps the simplest class of genuine blunt-body flows can be generated by a parabolic or paraboloidal shock wave placed symmetrically in a uniform free-stream of infinite Mach number. Because the free-stream Mach lines have zero slope, the parabolic and paraboloidal shocks have the correct asymptotic behaviour far downstream and thus they may be reasonable approximations to the shock waves produced by actual bodies during the early stages of atmospheric re-entry. Because of its relative simplicity, the axisymmetric paraboloidal shock-wave especially, is often used as a test case for numerical solutions.

While a series solution to the blunt-body problem in two space dimensions will, in general, require double Taylor series expansions for the dependent variables, this was not found to be necessary for the present cases. By using orthogonal coordinates it becomes possible to obtain results for each variable as a single Taylor series in the coordinate normal to the shock, whose coefficients may be computed recursively as exact functions of the other space coordinate. Thus, there is no drastic loss of accuracy away from the stagnation region; if it happens that the dependent variables are well-behaved functions throughout the entire shock layer, it should be possible to obtain an effectively exact solution to the entire problem rather than simply a good solution near the nose of the body. We will present results valid far downstream in the axisymmetric use. For plane flow, on the other hand, a limit line appears within the shock layer, suggesting the presence of an imbedded shock.

The problem may be formulated in orthogonal coordinates as in Fig. 3. For purposes of this section the angle of attack in the figure is always equal to zero. Taking the shock nose radius as unity, we introduce parabolic coordinates according to

$$x = \tfrac{1}{2}\,[1 + \xi^2 - (1 + \varepsilon)^2]$$
$$r = |\xi|\ (1 + \varepsilon). \tag{10}$$

The shock is seen to correspond to $\varepsilon = 0$. Dimensionless variables may be formed using ρ_∞, V_∞, and $\rho_\infty V_\infty^2$ as the reference quantities. The field equations become continuity:

$$\{ \xi^{\nu}(1+\nu\varepsilon) [\xi^2+(1+\varepsilon)^2]^{\frac{1}{2}}\rho u \}_{\xi} + \{ \xi^{\nu}(1+\nu\varepsilon) [\xi^2+(1+\varepsilon)^2]^{\frac{1}{2}}\rho v \}_{\varepsilon} = 0 \tag{11a}$$

ξ - momentum:

$$uu_{\xi} - \frac{\xi v^2}{\xi^2+(1+\varepsilon)^2} + v\left(u_{\xi} + \frac{(1+\varepsilon)u}{\xi^2+(1+\varepsilon)^2}\right) + \tau p_{\xi} = 0 \tag{11b}$$

ε - momentum:

$$vv_{\varepsilon} - \frac{(1+\varepsilon)u^2}{\xi^2+(1+\varepsilon)^2} + u\left(v_{\xi} + \frac{\xi v}{\xi^2+(1+\varepsilon)^2}\right) + \tau p_{\varepsilon} = 0 \tag{11c}$$

entropy:

$$uS+vT = 0 \tag{11d}$$

where the additional variables τ,S, and T have been introduced to remove cubic products and hence reduce the number of nested DO-loops in the computer program. Here

$$\tau = 1/\rho \tag{11c}$$

$$S = \rho p_{\xi} - \gamma p \rho_{\xi} \tag{11f}$$

$$T = \rho p_{\varepsilon} - \gamma p \rho_{\varepsilon}. \tag{11g}$$

ν=0 and 1 correspond to plane and axisymmetric flows, respectively. It can be shown that the system of equations (11) can be satisfied by assuming each independent variable to be an infinite series in ε where coefficients are polynomials in $\chi=1/(1+\xi^2)$. The density series, for example, is of the form

$$\rho(\varepsilon,\chi) = \sum_{j=0}^{\infty} \sum_{k=0}^{j} \rho_{jk}\, \varepsilon^{i}\, \chi^{k} \tag{12}$$

and thus the solution will be in the form of a triangular array of coefficients.

When expansions of the form (12) are substituted in (11), the differential equations reduce to algebraic relations where the unknown coefficients at each stage are computed recursively as sums of quadratic products of coefficients of lower order. A series solution for the stream function is developed from the series for ρ and u by term-by-term integration of

$$\psi_{\varepsilon} = \xi^{\nu}(1+\nu\varepsilon) [\xi^2+(1+\varepsilon)^2]^{\frac{1}{2}}\rho u \tag{13}$$

Additional details of the solution procedure may be found in reference 17. Power series solutions for both the plane and axisymmetric cases with γ=1.4 were found to $O(\varepsilon^{24})$ in about 1 minute of computer time for each case on the IBM 360/67. The coefficients for the first few orders may be recognised as rational numbers from their repeating decimals. For plane flow, we obtain for the stream function

$$\begin{aligned}\frac{\psi}{\xi} = 1 &+ 6\varepsilon + (45-35\chi)\varepsilon^2 + [330-(1355/3)\chi + 140\chi^2]\,\varepsilon^3 \\ &+ [4965/2 - (19055/4)\chi + (24175/9)\chi^2 - 455\chi^3]\,\varepsilon^4 \\ &+ [18648 - (92153/2)\chi + (681137/18)\chi^2 - (1304347/108)\chi^3+1316\chi^4]\,\varepsilon^5+\dots\end{aligned} \tag{14}$$

Similar expressions for the axisymmetric case were calculated to $O(\varepsilon^3)$ by Van Dyke[18] and a fourth term was added by Moran[7].

The results which follow were computed with [12/12] Padé approximants formed from the 24th order solution. The degree of convergence of the Padé table is excellent; the body (given by the streamline $\psi=0$) can be found to 10 decimal-place accuracy throughout the subsonic region for the axisymmetric case. The Padé table converges less well for the plane case. The stagnation point could be located to only 8 place accuracy. The convergence drops off dramatically as one enters the supersonic region, however, for this case. As an additional check, the pressure at the stagnation point, computed as a [12/12] approximant, agrees with the exact value to 11 and 7 places respectively for the two cases. Two separate methods were used to locate the body from the solution for ψ. The first method involves factoring the numerator of the Padé approximant for $\psi(\varepsilon;\chi)$ and identifying the body as the leading negative real zero for various constant values of χ. The body may also be found by reverting the series

$$\psi(\varepsilon;\chi) = \psi_0(\chi) + \psi_1(\chi)\varepsilon + \psi_2(\chi)\varepsilon^2+\ldots \tag{15}$$

to obtain

$$\varepsilon(\psi;\chi) = \varepsilon_1(\chi)(\psi-\psi_0) + \varepsilon_2(\chi)(\psi-\psi_0)^2+\ldots \tag{16}$$

and then recasting (16) as a Padé fraction with $\psi=0$. The accuracy of a solution of given order is about the same for the two methods. The series reversion method, however, is more efficient computationally and is more convenient for drawing streamlines.

Some significant features of the results are shown in Figures (4) through (7). Figure (4) shows the body shape and flow-field which support a parabolic shock for $\gamma=7/5$. The body shape was found from the expansion for the stream function, as were the typical streamlines in the figure. Note that the streamline $\psi=-.1$ lies "within" the body, i.e. in the analytic continuation of the shock-layer flow field.

The upstream limit line, corresponding to an envelope of characteristics as in the Taylor-Maccoll problem, and the continuation of the streamline $\psi=.3$ ahead of the shock are latent in the analytic solution which is, of course, "unaware" that the shock represents a physical discontinuity. As expected, the upstream limit line lies closer to the shock than does the body which indicates that the various series expansions will be divergent in the vicinity of the body. Also shown is the sonic line and its upstream and downstream analytic continuations. On this line the local Mach number, given as

$$M_\ell^2 = \frac{\rho}{\gamma p} - \frac{2}{\gamma-1} \tag{17}$$

assumes the value unity. For the plane case, the sonic line is clearly a closed curve, touching the upstream limit line on the line of symmetry. The surprising feature of Figure (4) is the presence of a limit line within the shock layer. On this line the density and other flow variables have infinite gradients. Unlike the upstream limit line, which exists only in a fictitious flow region, this limit line lies within the physically important region between the bow-shock and the body. Thus, in order to find that portion of the body lying behind the downstream limit line, it will be necessary to permit discontinuities of the field variables in the shock layer. A secondary, or embedded, shock might be inserted just upstream of the limit line and the solution could then be continued from the downstream side of this shock. It is doubtful, however, that such a modification is unique. It appears clear that no two-dimensional body can be found that will possess a parabolic bow-shock and have a flow field free of other discontinuities.

Figure 5 shows the body pressure distribution for the plane case. The circles in both figures 4 and 5 represent a numerical solution obtain with the marching-from-the-shock technique of Lomax and Inouye[19]. Their solution is graphically indistinguishable from the high order Padé - fraction results. The approximate results using only [2/2] fractions are indicated as dashed lines. This reasonably good agreement indicates that the earliest attempts to solve the blunt-body problem, using only hand-calculated series, might have been successful had the knowledge of the upstream limiting envelope existed at the time. Similar results for the axisymmetric case[17] show even closer agreement.

The two limit lines shown in Figure 4 were found through the use of Domb-Sykes plots. Three such plots for the density series, all indicating upstream singularities, appear in Figure 6. For $\xi=0$, corresponding to the line of symmetry, the plot clearly indicates a square-root singularity at the critical value $\varepsilon_+^* = .120$. For $\xi=0.5$, the asymptote has been drawn so as to indicate a square-root though other exponents are surely possible. Because of the uncertainty latent in any graphical extrapolation, the results of the third plot, for $\xi=\infty$, are particularly reassuring. Here the singularity exponent, obtained as a best-fit to the plotted points, is $\alpha=-1.88$. It was later determined that the system of equations (11) simplifies considerably in the limit $\xi\to\infty$ and that the reduced system possesses an exact local solution near a singular point ε^* of the form

$$\rho \sim (\varepsilon-\varepsilon^*)^{-3/(3-\gamma)}$$

For $\gamma=7/5$, the exponent is exactly equal to $-15/8=-1.875$. The inner limit line was found using Domb-Sykes plots after suitable Euler transformations were performed.

The axisymmetric case, unlike the two-dimensional flow, appears to possess a completely analytic shock layer. It is therefore possible to compute the body shape to great distances downstream. Figure 7 shows the afterbody computed to 200 shock nose radii downstream. It is in substantial agreement with Yakura's[20] asymptotic solution

$$\frac{r_b}{R_s} = 1.392 \left(\frac{x}{R_s} \right)^{\frac{1}{2\gamma}} \tag{18}$$

which he obtained by modifying the blast-wave analogy to account for the entropy layer. A similar but higher-order numerical solution of Sychev[21] is also shown. Our computed afterbody agrees well with Sychev's although the convergence of the approximants is less good past $x/R_s \simeq 100$. This loss of convergence is related to our use of parabolic coordinates; our body degenerates to the branch cut $\varepsilon=-1$ in the downstream limit.

IV Plane blunt-body flow at arbitrary Mach number and angle of attack

The method of Section III has been substantially generalized to treat both finite free-stream Mach numbers as well as arbitrary angles of incidence. While numerical methods have been largely successful in the treatment of symmetric configurations, it is only quite recently that, with the use of more general finite difference techniques and immensely more powerful computers, that promising solutions to the more general asymmetric problems have been solved even to engineering accuracy[22]. While the method of the present section treats only two-dimensional configurations, it produces solutions of very great accuracy that can serve as useful test-case comparisons for finite difference solutions. Moreover, no numerical solution appears to be of sufficient accuracy to adequately resolve the question of whether the maximum-entropy streamline wets the body surface in asymmetric flows. In this section we show that the stagnation streamline is displaced from the maximum entropy streamline by a relatively small, but by no means negligible amount. We also continue the discussion of limit lines and the domain of validity of the inverse method of solution.

Following Van Dyke and Gordon,[23] the bow-shock is described by

$$y^2 = 2R_s x - Bx^2. \tag{19}$$

Here B, the so-called shock bluntness, is a measure of the eccentricity. B=0 represents a parabola and B>1, an oblate ellipse, for example.

An orthogonal coordinate system with the shock corresponding to $\varepsilon=0$ is introduced by setting

$$x = \frac{1}{B}\{1 - [(1-B\xi^2)(1 + 2B\varepsilon + B\varepsilon^2)]^{\frac{1}{2}}\} \tag{20a}$$

and

$$y = \xi(1 + \varepsilon) \tag{20b}$$

which is a generalization of the transformation (10) of the last section.

The initial conditions are specified at the shock via the Rankine-Hugoniot relations:

$$\rho\,[0,\xi(\theta)] = \frac{(\gamma+1)M_\infty^2 \sin^2\theta}{(\gamma-1)M_\infty^2 \sin^2\theta+2} \tag{21a}$$

$$p\,[0,\xi(\theta)] = \frac{2}{\gamma+1}\sin^2\theta - \frac{(\gamma-1)}{(\gamma+1)\gamma M_\infty^2} \tag{21b}$$

$$u\,[0,\xi(\theta)] = \cos\theta \tag{21c}$$

$$v\,[0,\xi(\theta)] = \frac{(\gamma-1)M_\infty^2 \sin^2\theta+2}{(\gamma+1)M_\infty^2 \sin^2\theta}\sin\theta. \tag{21d}$$

Here the uniform free stream of Mach number M_∞ is inclined at an angle α to the x-axis as in Figure 3 and θ is given by

$$\cos\theta = \frac{\xi \cos\alpha + (1-B\xi^2)^{\frac{1}{2}} \sin\alpha}{[1+(1-B)\xi^2]^{\frac{1}{2}}} . \tag{22}$$

Equations (21) are the initial conditions for the gasdynamic equations which, in (ε,ξ) coordinates, become

$$R^{(2)}(\xi)\left\{[C\xi^2 + (1+\varepsilon)^2](\rho u)_\xi + C\xi \rho u\right\} + R^{(1)}(\varepsilon)\left\{[C\xi^2 + (1+\varepsilon)^2](\rho v)_\varepsilon + (1+\varepsilon)\rho v\right\} = 0, \tag{23a}$$

$$R^{(2)}(\xi)\left\{[uu_\xi + \tau p_\xi][C\xi^2 + (1+\varepsilon)^2] - C\xi v^2\right\} + R^{(1)}(\varepsilon)\left\{vu_\varepsilon[C\xi^2 + (1+\varepsilon)^2] + (1+\varepsilon)uv\right\} = 0, \tag{23b}$$

$$R^{(2)}(\xi)\left\{uv_\xi[C\xi^2 + (1+\varepsilon)^2] + C\xi uv\right\} + R^{(1)}(\varepsilon)\left\{[vv_\varepsilon + \tau p_\varepsilon][C\xi^2 + (1+\varepsilon)^2] - (1+\varepsilon)u^2\right\} = 0, \tag{23c}$$

and

$$R^{(2)}(\xi)[u(\rho p_\xi - \gamma p \rho_\xi)] + R^{(1)}(\varepsilon)[v(\rho p_\varepsilon - \gamma p \rho_\varepsilon)] = 0. \tag{23d}$$

Here

$C=1-B$,

$R^{(1)}(\varepsilon) = [1+B(2\varepsilon+\varepsilon^2)]^{\frac{1}{2}}$,

and $R^{(2)}(\xi) = (1-B\xi^2)^{\frac{1}{2}}$.

While in the cases treated in the preceding section, it was possible to express the solution for each dependent variable as a single power series in ε with coefficients which are exact functions of ξ, in the present more general case, this was not practical. Here the series coefficients involve the parameters M_∞, B, and α as well as γ. Moreover, the dependence on ξ is sufficiently complex that, in the interest of computational efficiency, the single series must be abandoned in favour of a double power series solution. Thus, for the density, we assume an expansion of the form

$$\rho(\varepsilon,\xi) = \sum_{j=0}^{\infty} \sum_{k=0}^{\infty} \rho_{jk}\, \varepsilon^j \xi^k \tag{24}$$

with similar expressions for the other dependent variables. By inserting these expansions in the system (23), performing the series multiplications, and equating coefficients of terms of like order in both ε and ξ, recurrence relations may be derived for the elements ρ_{jk}, p_{jk}, etc. Additional details may be found in reference 24.

It is a feature of the recurrence procedure that the number of terms in ξ which may be found exactly at each stage decreases as the order of ε is increased. Thus if the computation is started by expanding the Rankine-Hugoniot relations (21) as a series in ξ up to a given order N, there will exist only a sufficient number of relations to determine the first-order coefficients in ε to order N-1 in ξ. Similarly, at the next stage, only terms up to $O(\varepsilon^2\xi^{N-2})$ may be found. The final result will be, therefore, a triangular array of coefficients, for each variable, which includes all elements of total order, that is the sum of the exponents of ε and ξ, $\leq N$.

A formal extension of the Padé concept to cases involving two independent variables may be effected by grouping all terms of the same total order together. Consider $f(\varepsilon,\xi)$ to be a typical dependent variable and form

$$\begin{aligned} f(\varepsilon,\xi) &= \sum_{i=0}^{\infty}\sum_{j=0}^{\infty} f_{ij}\,\varepsilon^i\xi^j = \sum_{k=0}^{\infty}\sum_{j=0}^{k} f_{k-j,j}\,\varepsilon^{k-j}\xi^j \\ &= \sum_{k=0}^{\infty} \varepsilon^k\left[\sum_{j=0}^{k} f_{k-j,j}\left(\frac{\xi}{\varepsilon}\right)^j\right] \end{aligned} \tag{25}$$

which is a single power series in ε whose coefficients are polynomials in (ξ/ε). Standard methods for treating single series, e.g. Wynn's epsilon algorithm[25], may now be used. Note that this method is fully compatible with the triangular arrays produced by the recurrence formulas. Various other techniques for treating multiple power series have been devised by Chisholm[26] and his group.

The double power series, as derived, are expansions about the shock apex. Clearly, any other point on the shock could serve just as well. The maximum-entropy point, where the shock is normal to the free-stream, is perhaps a more logical choice and might result in more uniform convergence in the results which follow.

The body produced by a given shock is located by first finding the stagnation point, which is a saddle-point of the stream function. Starting with an initial guess, the desired condition u=v=0 may be approached via a Newton iteration:

$$\varepsilon_{i+1} = \varepsilon_i + \frac{(vu_\xi - uv_\xi)_i}{(u_\varepsilon v_\xi - v_\varepsilon u_\xi)_i} \tag{26a}$$

$$\xi_{i+1} = \xi_i + \frac{(uv_\varepsilon - vu_\varepsilon)_i}{(u_\varepsilon v_\xi - v_\varepsilon u_\xi)_i} \quad . \tag{26b}$$

Note that the partial derivatives need not be evaluated numerically but may be found by term-by-term differentiation of the power series which is then recast as a Padé fraction. The body shape can then be found by finding points which lie on the streamline passing through the stagnation point. Other streamlines as well as the sonic lines can be found using Newton iteration.

Typical results are shown in Figures 8 and 9. Various total orders of solution are shown for each case. A parabolic shock is set at 10^0 incidence to a free-stream with $M_\infty = 2$ in Figure 8. The 30th order solution, computed with [15/15] approximants and the procedure of (25), is sensibly exact. Notice that the maximum entropy streamline passes below the streamline that wets the body surface. Since the flow-field is rotational, the stagnation streamline does not intercept the body exactly at right angles. It exhibits a small characteristic bending away from the maximum entropy streamline. In Figure 9, a flat-faced $\sqrt{2}$-to-1 ellipse is set at 10^0 incidence with infinite free-stream Mach number. Here the maximum entropy streamline passes above the one that wets the body. The solution here also is fully converged by 30th order. A high order solution is required here to locate the sonic lines correctly. It was decided to plot the location of the limit lines for this case to see whether the failure of the low order solutions can be better understood. As before, Domb-Sykes plots are used, coupled with suitable Euler transformations when necessary. The limit lines have been mapped out in Figure 10. The limit lines appear both within the supersonic position of the shock layer and also in the "fictitious" upstream analytic continuation of the flow. The line opposite the shock exhibits a square-root singularity ($\delta=\frac{1}{2}$) as do the limit lines within the shock layer. These are augmented by singular lines with $\delta\simeq-0.43$ placed roughly symmetrically in the analytic continuation. Lines AB&AG can be traced back to point A on the shock. Here the shock slope = 10^0, the free-stream angle of attack. At this point the shock becomes a Mach line and hence it is plausible that it should be a singular point. Note how close the limit lines come to the sonic lines which explains the need for high order solutions in these regions. Because of the close proximity of the limit lines, obtaining a starting line for a method of characteristic solution would appear to be a practical impossibility for this case.

Various other features of this asymmetric blunt-body solution are explored in reference 24. Cases up to 30^0 angle-of-attack have been treated. The displacement of the maximum entropy streamline from the stagnation streamline appears to increase somewhat faster than linearly with angle of attack. It is also increased as the Mach number is lowered. The body pressure distribution can also be found to good accuracy.

The series expansion method would appear to merit further study as a possible alternative to finite-difference techniques for the asymmetric problem. Three dimensional solutions are also possible where the substantial algebra required to derive the recurrence relations for the series coefficients might be alleviated through the use of a symbol manipulation language such as FORMAC.

Acknowledgement

Most of this work was performed while the author was a post-doctoral associate at the NASA Ames Research Center. Financial support during this period was provided by the National Research Council.

References

1. J. S. Hadamard, Lectures on Cauchy's Problem in Linear Partial Differential Equations, (Yale Univ. Press, New Haven, 1923).

2. J. Dugundji, Jour. Aero. Sci. 15 699 (1948).
3. H. Cabannes, ONERA note technique no. 5 (1951).
4. C. C. Lin and S. F. Shen, NACA tech. Note 2506 (1951).
5. M. D. Van Dyke, J. Fluid Mech. 3, 515 (1958).
6. A. Van Tuyl, J. Aero. Sci. 27 559 (1960).
7. J. P. Moran, Ph.D. Thesis, Cornell Univ. Ithaca, New York (1965).
8. C. Domb and M. F. Sykes, Proc. Roy. Soc. (London) A240, 214 (1957).
9. M. D. Van Dyke, J. Fluid Mech. 44 365 (1970).
10. M. D. Van Dyke, Quart. J. Mech. Appl. Math. 27 423 (1974).
11. G. I. Taylor and J. W. Maccoll, Proc. Roy. Soc. London A139 279 (1933).
12. Z. Kopal, Tech. Rept. No. 1. Mass. Inst. Tech. (1974).
13. J. L. Sims, NASA SP-3004 (1964).
14. J. L. Sims, NASA SP-3078 (1973).
15. J. W. Maccoll, Proc. Roy.Soc. London A159 459 (1937).
16. L. W. Schwartz, Jour. Appl. Math. Phys. (ZAMP) to appear 26 (1975).
17. L. W. Schwartz, Phys. of Fluids 17, 1816 (1974).
18. M. D. Van Dyke in Fundamental Phenomena in Hypersonic Flow (Cornell Univ. Press, Ithaca, New York, 1966) p.52.
19. H. Lomax and M. Inouye, NASA Tech. Report R204 (1964).
20. J. K. Yakura, in Hypersonic Flow Research (Academic, New York, 1962) p.241.
21. V. V. Sychev, Prik. Mat. Mekh. 24, 518 (1960) [J. Appl. Math. Mech. 24 756 (1960)] .
22. A. Rizzi and M. Inouye, A.I.A.A. J. 11, 1478 (1973).
23. M. D. Van Dyke and H. D. Gordon, NASA Tech. Rept. R-1 (1959).
24. L. W. Schwartz, Physics of Fluids to appear 1975.
25. P. Wynn, SIAM J. Numer. Anal. 3,91 (1966).
26. J. S. R. Chisholm, Math. Comp. 27, 841 (1973).

LIST OF FIGURE CAPTIONS

Figure 1 Domb-Sykes plot for radial velocity series coefficients (θ_c=40^0, M_c=2, γ=7/5).

Figure 2 Radial velocity component (θ_c=5^0, γ=7/5)

Figure 3 Blunt-body coordinate system.

Figure 4 Flow field, parabolic shock, M_∞=∞, γ=7/5.

Figure 5 Body pressure distribution, parabolic shock.

Figure 6 Domb-Sykes plot of density series, upstream limit line, parabolic shock.

Figure 7 Afterbody coordinates, paraboloidal shock, γ=7/5.

Figure 8 Flow field, parabolic shock, M_∞=2, α=10^0, γ=7/5, ooo N=20, □ □ □ N=26, ——N=30, --- Maximum entropy streamline.

Figure 9 Flow field, elliptic shock, M_∞=∞, α=10^0, γ=7/5.

Figure 10 Location of limit lines for the flow field of Figure 9.

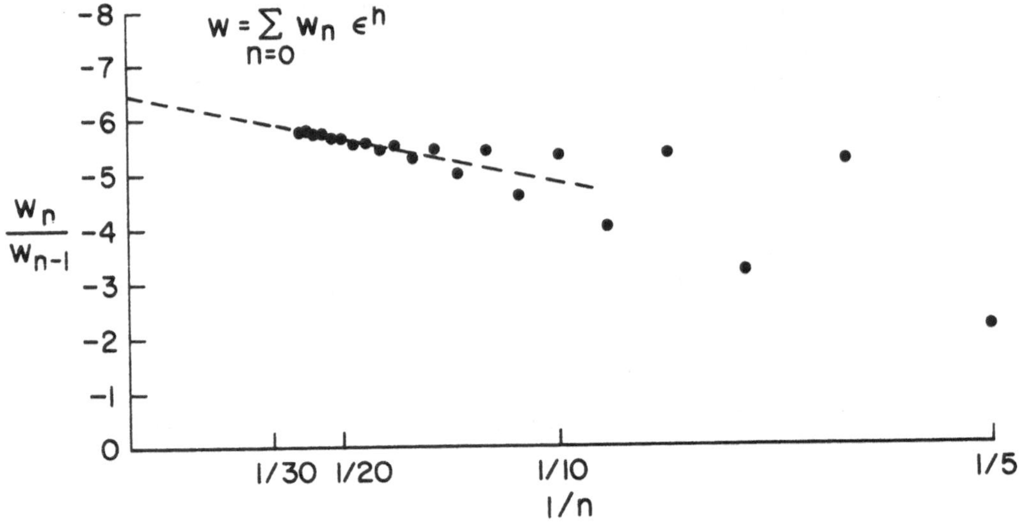

-8
-7
-6
-5
-4
-3
-2
-1
0
$w = \sum_{n=0} w_n \epsilon^n$
$\frac{w_n}{w_{n-1}}$
1/30 1/20
1/10
1/5
1/n

Fig. 1

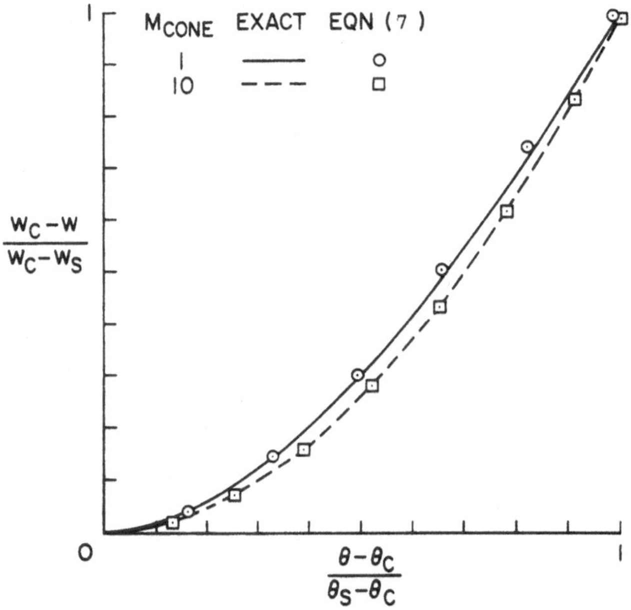

M_{CONE} EXACT EQN (7)
1
10
0
1
$\frac{w_C - w}{w_C - w_S}$
$\frac{\theta - \theta_C}{\theta_S - \theta_C}$

Fig. 2

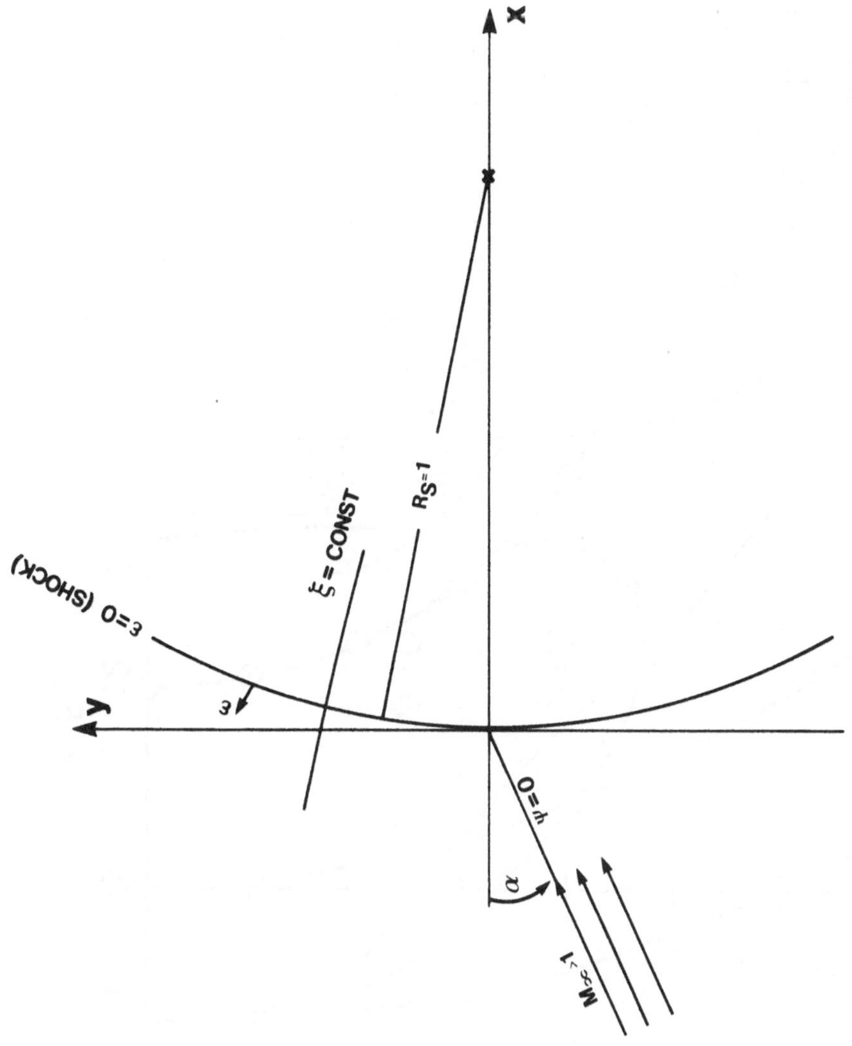

Fig. 3

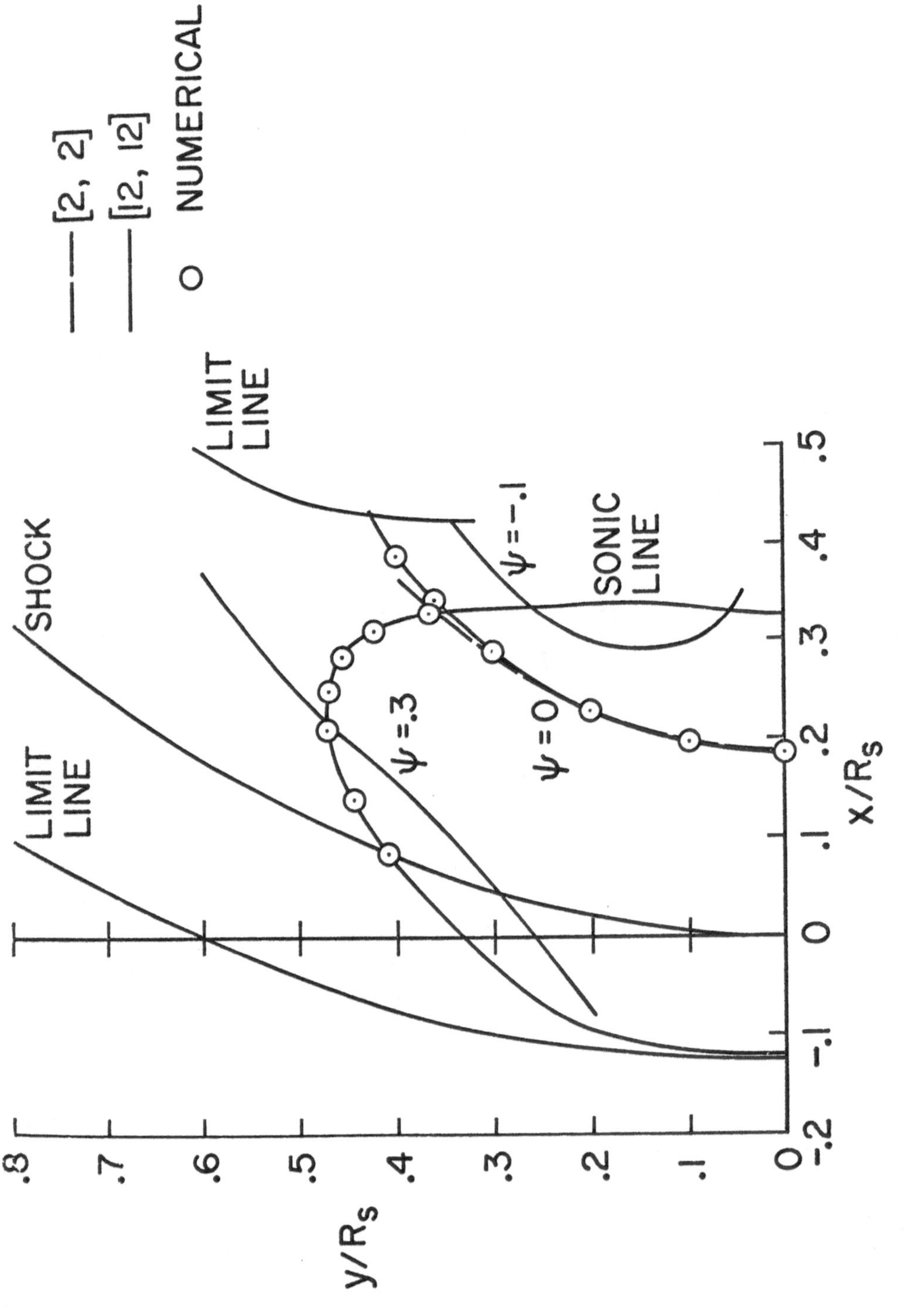

[2, 2]
[12, 12]
O NUMERICAL
LIMIT LINE
SHOCK
LIMIT LINE
SONIC LINE
ψ = −.1
ψ = .3
ψ = 0
y/R_s
x/R_s
.8
.7
.6
.5
.4
.3
.2
.1
0
−.2
−.1
0
.1
.2
.3
.4
.5

Fig. 4

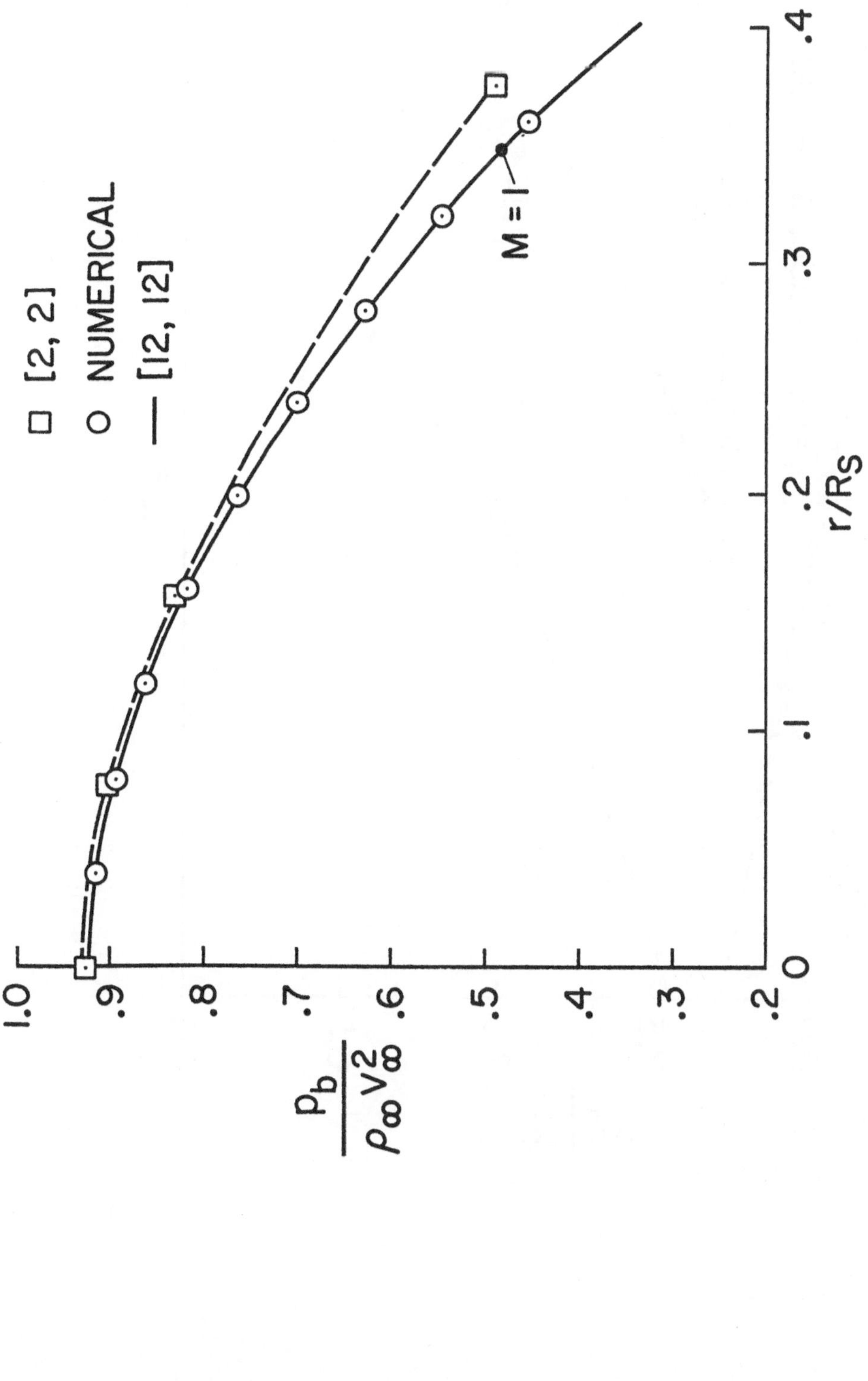

□ [2, 2]
○ NUMERICAL
— [12, 12]
M = 1
$\frac{p_b}{\rho_\infty V_\infty^2}$
1.0
.9
.8
.7
.6
.5
.4
.3
.2
0
.1
.2
.3
.4
r/R_S

Fig. 5

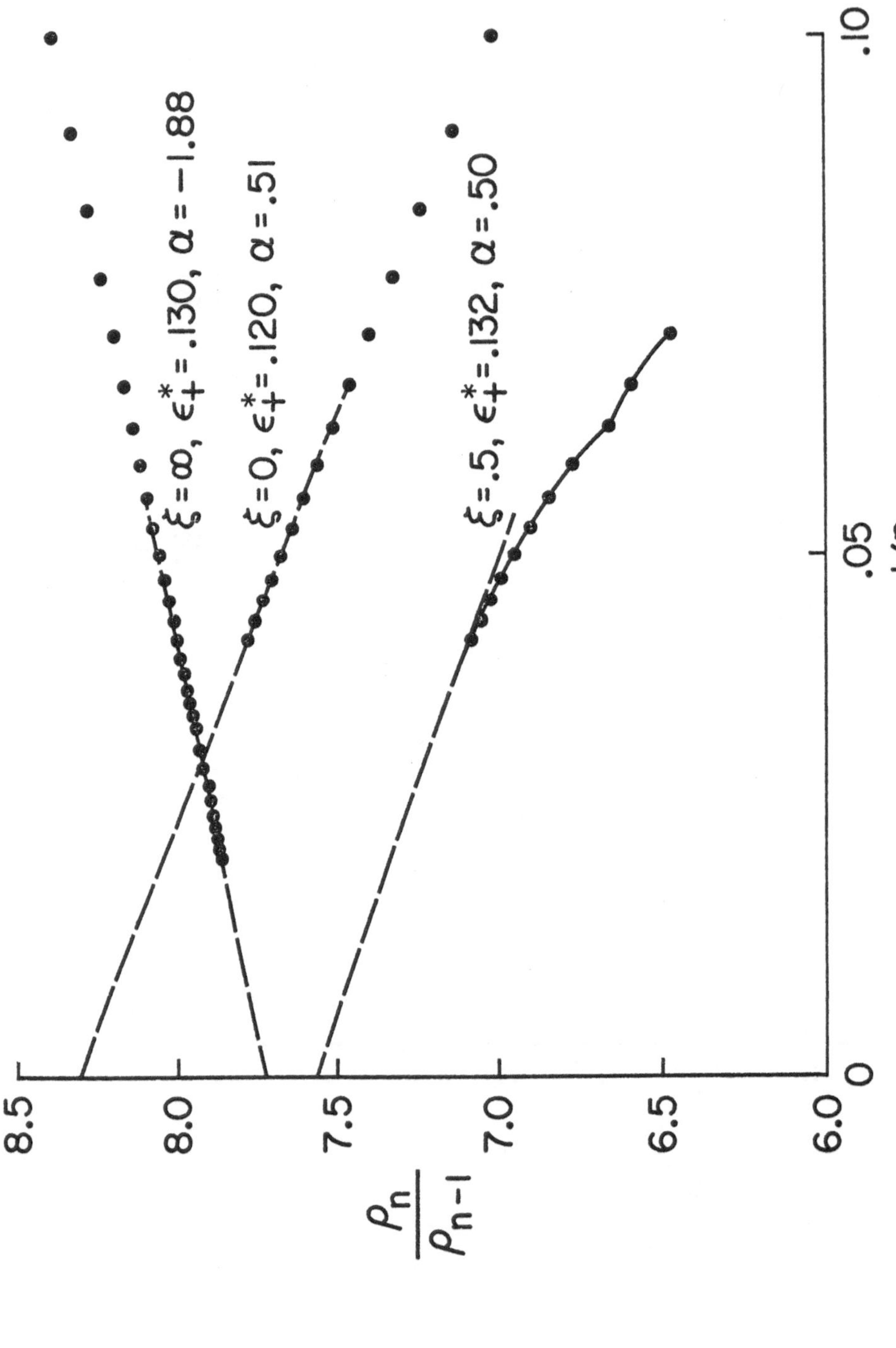

$\xi = \infty$, $\epsilon_+^* = .130$, $\alpha = -1.88$
$\xi = 0$, $\epsilon_+^* = .120$, $\alpha = .51$
$\xi = .5$, $\epsilon_+^* = .132$, $\alpha = .50$
$\frac{\rho_n}{\rho_{n-1}}$
8.5
8.0
7.5
7.0
6.5
6.0
0
.05
.10
1/n

Fig. 6

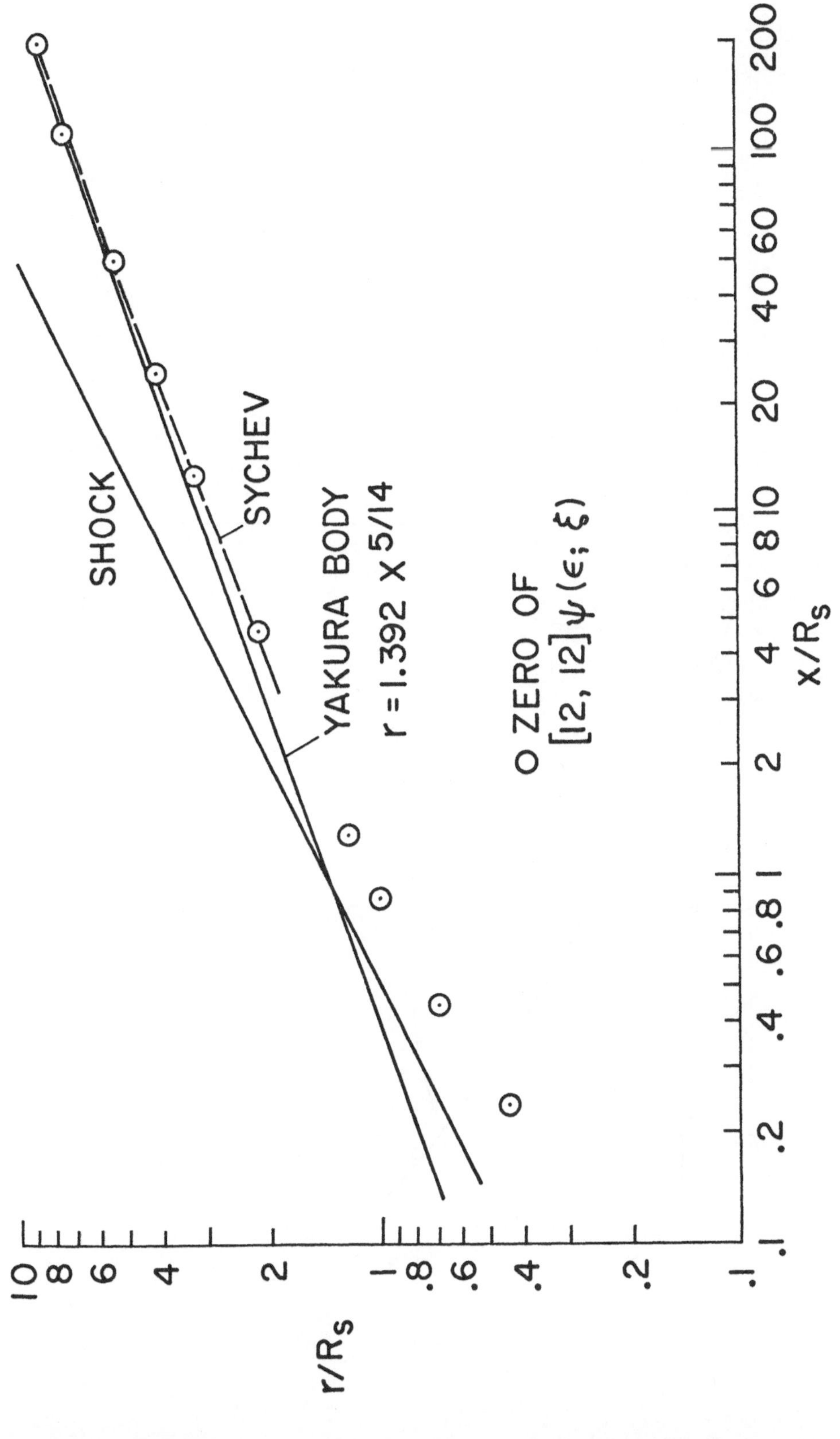
SHOCK
SYCHEV
YAKURA BODY
r = 1.392 X 5/14
O ZERO OF
[12, 12] ψ(ε; ξ)
x/R_S
r/R_S

Fig. 7

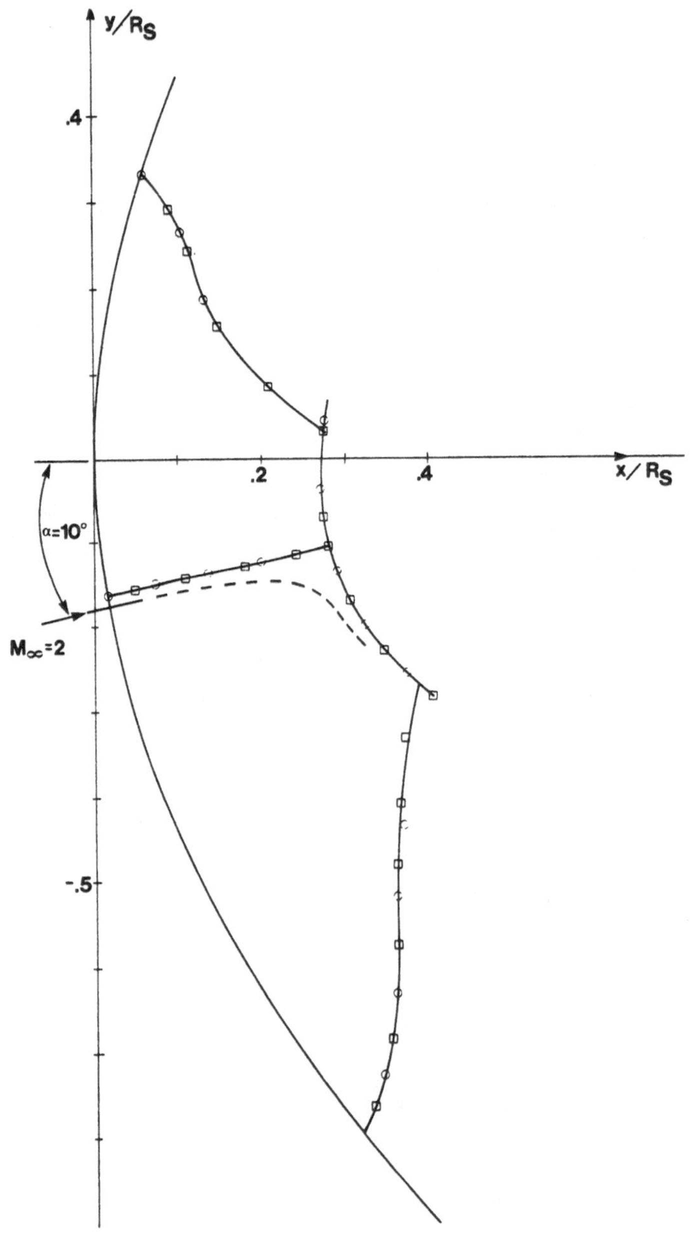

Fig. 8

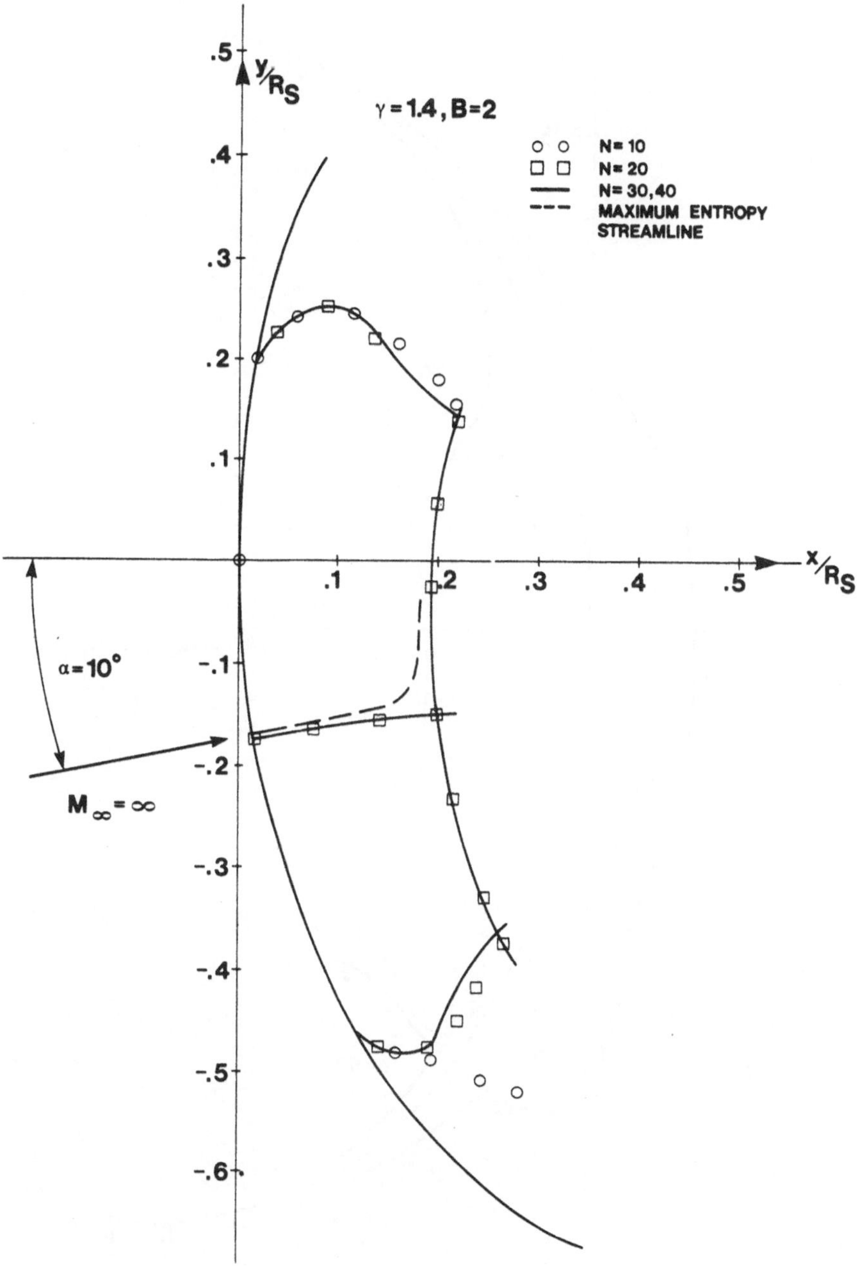

.5
y/RS
γ = 1.4, B=2
.4
.3
.2
.1
○ ○ N= 10
□ □ N= 20
N= 30,40
MAXIMUM ENTROPY STREAMLINE
.1
.2
.3
.4
.5
x/RS
α=10°
-.1
-.2
-.3
-.4
-.5
-.6
M∞ = ∞

Fig. 9

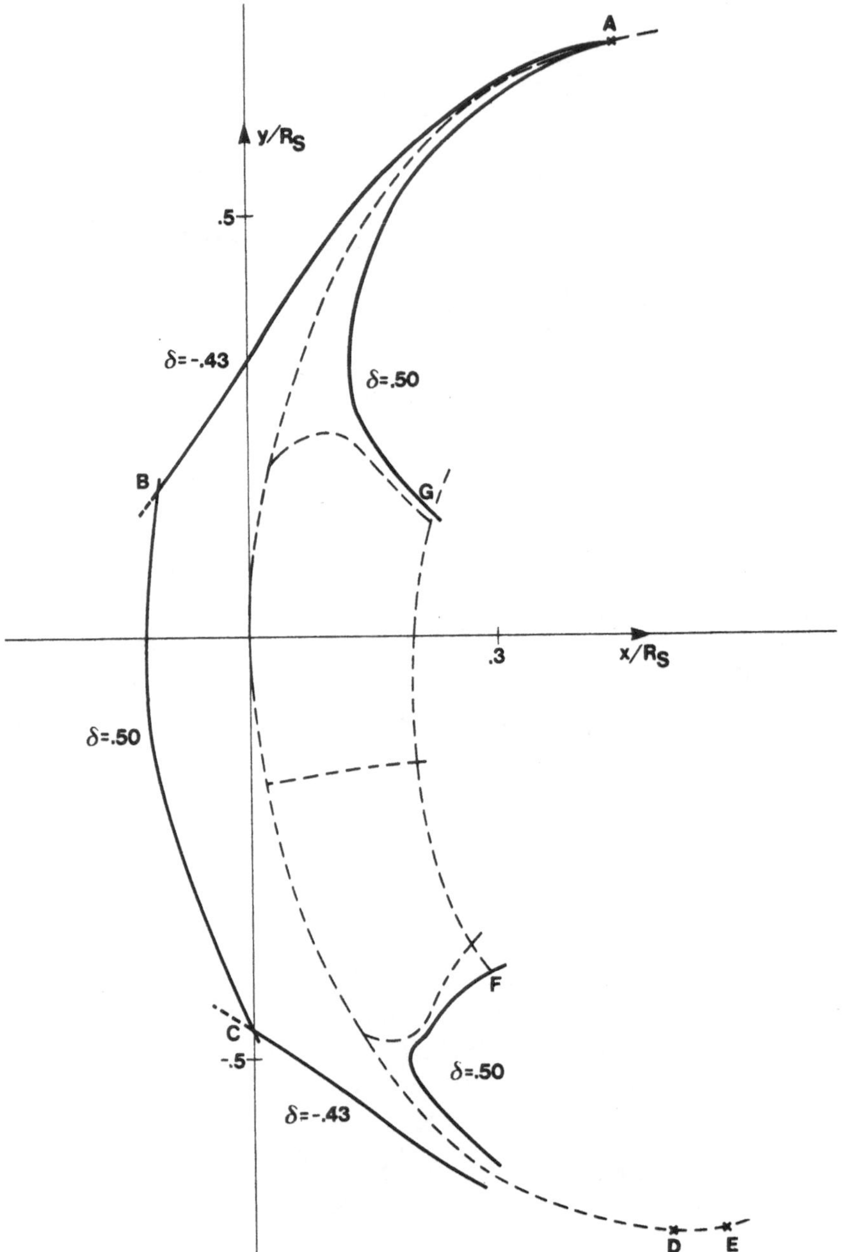

Fig. 10

WAVE FRONT EXPANSIONS AND PADE' APPROXIMANTS FOR TRANSIENT WAVES IN LINEAR DISPERSIVE MEDIA

G. Turchetti, F. Mainardi

Istituto di Fisica, Università di Bologna, Italy
Gruppo Nazionale per la Fisica Matematica del C.N.R.

1. INTRODUCTION

The wave propagation in linear dispersive media has different and interesting aspects. Many of them concern plane waves propagating in a homogeneous half space subjected to a known input condition at the free surface. The related boundary-value problem is conveniently treated by the Laplace transform. The transform solution is easily determined, however its inversion is a very difficult task except for some special cases. Usually short and long time approximations are derived in the literature.

In this note we review a simple method which enables us, using the Padé Approximants, to compute the solutions of most transient wave problems in space-time domains of physical interest. Our analysis is based on recent investigations about viscoelastic and thermoelastic waves |1, 2|, with a particular emphasis on the numerical aspects.

The plane of the work is the following.

In Sections 2. and 3. we consider the wave propagation in linear viscoelastic and thermoelastic media, which leads to the same transform equation.

In Section 4. we derive a wave front expansion which provides a series solution, uniformly convergent in any space-time domain.

In Section 5. we discuss the numerical properties of this solution and introduce the diagonal Padé Approximants in order to accelerate the convergence.

In Section 6. we present some results which confirm the efficiency of the method.

2. VISCOELASTIC WAVES

The basic equations of the dynamic theory of linear viscoelasticity are, in the unidimensional case (see for example |3,4,5|):

$$\frac{\partial}{\partial x}\sigma(x,t) = \varrho\frac{\partial^2}{\partial t^2}u(x,t) \tag{2.1}$$

$$\varepsilon(x,t) = \left\{J_0 + \frac{d}{dt}J(t)*\right\}\sigma(x,t) \tag{2.2}$$

$$\varepsilon(x,t) = \frac{\partial}{\partial x}u(x,t) \tag{2.3}$$

$$v(x,t) = \frac{\partial}{\partial t}u(x,t) \tag{2.4}$$

where σ is the stress, ε the strain, u the displacement, v the particle velocity, ϱ the density, $J(t)$ the creep compliance, $J_0 = J(0^+)$, and $*$ denotes the Riemann convolution.

Denoting with R (response variable) any one of the variables σ, ε, u, v, we consider the problem of determining $R=R(x,t)$ with the initial conditions:

$$R(x,t) = \frac{\partial}{\partial t}R(x,t) = 0 \qquad \text{for} \qquad t=0 \tag{2.5}$$

and the boundary conditions:

$$\begin{aligned} R(x,t) &= R_0(t) && \text{for} \quad x=0 \\ R(x,t) &\to 0 && \text{as} \quad x\to\infty \end{aligned} \tag{2.6}$$

Taking the Laplace transform of equations (2.1) - (2.5) we obtain for the transform response $\bar{R}(x,s)$ the following differential equation:

$$\left\{\frac{\partial^2}{\partial x^2} - \mu^2(s)\right\}\bar{R}(x,s) = 0 \tag{2.7}$$

where

$$\mu^2(s) = \frac{s^2}{c^2}\left[1+\bar{\psi}(s)\right] \tag{2.8}$$

In (2.8) $c=(\varrho J_0)^{-\frac{1}{2}}$ and $\bar{\psi}(s)$ is the Laplace transform of the rate of creep $\psi(t) = \frac{1}{J_0}\frac{d}{dt}J(t)$.

Accounting for (2.6) the solution of (2.7) reads:

$$\bar{R}(x,s) = \bar{R}_0(s)\exp[-x\,\mu(s)] \tag{2.9}$$

with $\mu(s) \geq 0$ for $\arg(s)=0$ |3|.

In most cases $\psi(t)$ is an entire function of exponential type so that $\bar{\psi}(s)$ is analytic and vanishing at infinity, as known from the transform theory |6|. This occurs for viscoelastic models whose relaxa tion spectrum is strictly positive. Then $\mu(s)$ is an analytic function in the s-plane cut along a finite domain of the negative real axis |1|.

A relevant model of viscoelasticity is the Standard Linear Solid (SLS) for which:

$$\psi(t) = (1 - a)\, e^{-at} \Rightarrow \bar{\psi}(s) = (1 - a)/(s + a) \tag{2.10}$$

where 1 and $1/a$ represent the relaxation time and the retardation time respectively |3|. For $a = 0$ we recover the Maxwell Solid, for which the wave equation reduces to the 'telegraph equation' |7|.

3. THERMOELASTIC WAVES

A generalized dynamic theory of linear thermoelasticity has been introduced by several authors (see for example |8,9,10,11|)in order to eliminate the paradox, present in the classical theory, of an infinite velocity for thermal and mechanical disturbances. In the unidimensional case the basic equations are, in a convenient non-dimensional form, |10|

$$\frac{\partial}{\partial x} S(x,t) = \frac{\partial^2}{\partial t^2} U(x,t) \tag{3.1}$$

$$S(x,t) = E(x,t) - \Theta(x,t) \tag{3.2}$$

$$E(x,t) = \frac{\partial}{\partial x} U(x,t) \tag{3.3}$$

$$\frac{\partial^2}{\partial x^2} \Theta(x,t) - \frac{\partial}{\partial t} \Theta(x,t) - \beta \frac{\partial^2 \Theta}{\partial t^2} = \varepsilon \left[\frac{\partial}{\partial t} E(x,t) + \beta \frac{\partial^2}{\partial t^2} E(x,t)\right] \tag{3.4}$$

where S is the stress, E the strain, U the displacement, Θ the temperature, ε the thermoelastic coupling constant and β the relaxation constant introduced to account for the acceleration of the heat flux.

We consider the problem of determining the temperature $\Theta = \Theta(x,t)$ and the strain $E = E(x,t)$ with the initial conditions:

$$\Theta(x,t) = \frac{\partial}{\partial t} \Theta(x,t) = E(x,t) = \frac{\partial}{\partial t} E(x,t) = 0 \qquad \text{for} \qquad t = 0 \tag{3.5}$$

and the boundary conditions:

$$\Theta(x,t) = \Theta_o(t)\ , \quad E(x,t) = E_o(t) \qquad \text{for} \qquad x = 0$$
$$\Theta(x,t) \to 0\ , \quad E(x,t) \to 0 \qquad \text{as} \qquad x \to \infty \tag{3.6}$$

Taking the Laplace transforms of equations (3.1) - (3.5) we obtain for the transform responses $\bar{\Theta}(x,s)$, $\bar{E}(x,s)$ two coupled differential equations. However it is possible to uncouple the problem |2| by setting

$$\bar{\Theta} = \bar{\Theta}^{+} + \bar{\Theta}^{-} \qquad \bar{E} = \bar{E}^{+} + \bar{E}^{-} \tag{3.7}$$

where $\bar{\Theta}^{\pm}$, $\bar{E}^{\pm}$ satisfy the following equations:

$$\left\{\frac{\partial^2}{\partial x^2} - \mu_{\pm}^2(s)\right\} \bar{\Theta}^{\pm}(x,s) = 0$$
$$\left\{\frac{\partial^2}{\partial x^2} - \mu_{\pm}^2(s)\right\} \bar{E}^{\pm}(x,s) = 0 \tag{3.8}$$

In (3.8) $\mu_{+}^2(s)$, $\mu_{-}^2(s)$ are the algebraic roots of the equation:

$$\mu^4 - s^2 u \mu^2 + s^4 v = 0 \tag{3.9}$$

with $u = 1 + \beta(1+\varepsilon) + (1+\varepsilon)/s$, $v = \beta + 1/s$.

It is not difficult to prove that:

$$\mu_{\pm}^2(s) = \frac{s^2}{c_{\pm}^2}[1 + \bar{\psi}_{\pm}(s)] \tag{3.10}$$

where $c_{\pm} = \{[1 + \beta(1+\varepsilon) \pm \gamma]/2\}^{-\frac{1}{2}}$ with $\gamma = \{[1 + \beta(1+\varepsilon)]^2 - 4\beta\}^{\frac{1}{2}}$ and $\bar{\psi}_{\pm}(s)$ are functions analytic and vanishing at infinity. We remark that $\mu_{+}^2 \leftrightarrow \mu_{-}^2$ for $\gamma \leftrightarrow -\gamma$.

Accounting for (3.6) the partial solutions read:

$$\bar{\Theta}^{\pm}(x,s) = \left\{a_{11}^{\pm}(s)\ \bar{\Theta}_o(s) + a_{12}^{\pm}(s)\ \bar{E}_o(s)\right\} \exp[-x\ \mu_{\pm}(s)]$$
$$\bar{E}^{\pm}(x,s) = \left\{a_{21}^{\pm}(s)\ \bar{\Theta}_o(s) + a_{22}^{\pm}(s)\ \bar{E}_o(s)\right\} \exp[-x\ \mu_{\pm}(s)] \tag{3.11}$$

where $a_{ik}(s)$ $(i,k=1,2)$ are known functions and $\mu_{\pm}(s) \geq 0$ for $\arg(s) = 0$. The $a_{ik}(s)$ prove to be analytic at infinity with a non-vanishing value, and the $\mu_{\pm}(s)$ are analytic in the s-plane cut along the negative real axis between $s = -1/\beta$ and $s = 0$ |2| .

4. WAVE FRONT EXPANSIONS

In the previous Sections we have shown that the problem of transient waves can be reduced to the inversion of the Laplace transform:

$$\bar{R}(x,s) = \bar{f}(s)\ a(s)\exp\left\{-\frac{xs}{c}\left[1+\bar{\Psi}(s)\right]^{\frac{1}{2}}\right\} \tag{4.1}$$

where $\bar{f}(s)$ is the transform of a given input function $f(t)$, and $a(s)$, $\bar{\Psi}(s)$ are some analytic functions regular at infinity, with $a(\infty) \neq 0$, $\bar{\Psi}(\infty) = 0$ ($a(s) = 1$ for the viscoelastic case).

The inversion of (4.1) is simpler when $\bar{f}(s) = 1$. The corresponding function, denoted by $G(x,t)$ (Green's function), enables us to obtain $R(x,t)$ for any input $f(t)$ by the Riemann convolution:

$$R(x,t) = f(t) * G(x,t) \tag{4.2}$$

The wave properties of $R(x,t)$ appear from the limit as $s \to \infty$ of $\bar{G}(x,s)$, which provides explicitly the discontinuity and the velocity of the wave front and the space damping. From the previous considerations we can expand $a(s)$ and $\bar{\psi}(s)$ in Laurent series, according to:

$$a(s) = a_0 + a_1/s + a_2/s^2 + \ldots, \quad a_0 \neq 0 \tag{4.3}$$

$$\bar{\Psi}(s) = \psi_1/s + \psi_2/s^2 + \ldots \tag{4.4}$$

so that the limit of $\bar{G}(x,s)$ as $s \to \infty$ is:

$$\bar{G}(x,s) \simeq a_0 \exp\left[-\frac{xs}{c} - \lambda x\right] \tag{4.5}$$

with

$$\lambda = \frac{\psi_1}{2c} \tag{4.6}$$

Then for $t \to (x/c)^+$ we get:

$$R(x,t) \simeq f(t - x/c)\ a_0 \exp\left[-\lambda x\right] \tag{4.7}$$

from which we recognize that c is the wave front velocity, λ the space damping and $f(0^+)\, a_0 \exp[-\lambda ct]$ the jump of R at the wave front. Of course the decay condition $\lambda > 0$ is insured in our cases.

By accounting for the first few terms in $1/s$ of $\bar{R}(x,s)$ which can be analytically computed, short time approximationsare usually derived. On the other hand long time approximations are deduced from (4.1) either by the saddle point method or by the limit as $s \to 0$.

Here we sketch a recursive method to generate the whole expansion of $\overline{G}(x,s)$ in powers of $1/s$, which provides a wave front expansion of the Green's function, uniformly convergent for any x. For this purpose we put:

$$\overline{G}(x,s) = \exp[-\frac{xs}{c}]\,\hat{G}(x,s) \tag{4.8}$$

where $\hat{G}(x,s)$, which is analytic at infinity, is to be expanded in Laurent series according to:

$$\hat{G}(x,s) = w_o(x) + \sum_{k=1}^{\infty} w_k(x)/s^k \tag{4.9}$$

To determine the functions $w_k(x)$ $(k=0,1,\ldots)$ we consider the differential equation satisfied by $\hat{G}(x,s)$, namely, after (2.7), (2.8), (4.1), (4.8),

$$\left\{ \frac{\partial^2}{\partial x^2} - \frac{2s}{c}\frac{\partial}{\partial x} - \frac{s^2}{c^2}\,\overline{\psi}(s) \right\} \hat{G}(x,s) = 0 \tag{4.10}$$

which is subjected to the boundary condition $\hat{G}(0,s) = a(s)$. Inserting the expansions (4.3), (4.4), (4.9) into equation (4.10) and collecting like powers of s, we obtain the following recursive system of first order differential equations:

$$\begin{cases} \dfrac{d}{dx} w_o + \dfrac{\psi_1}{2c} w_o = 0 \\ \dfrac{d}{dx} w_k + \dfrac{\psi_1}{2c} w_k = \dfrac{c}{2}\dfrac{d^2}{dx^2} w_{k-1} - \dfrac{1}{2c}\displaystyle\sum_{J=1}^{k} \psi_{J+1}\, w_{k-J} \end{cases} \tag{4.11}$$

with initial conditions $w_k(0) = a_k$ $(k=0,1,2,\ldots)$. It is not difficult to prove that the $w_k(x)$ can be expressed by:

$$(k=0,1,2,\ldots) \qquad w_k(x) = \exp(-\lambda x) \sum_{h=0}^{k} A_{kh} \frac{x^h}{h!} \tag{4.12}$$

where λ is given by (4.6) and the A_{kh} are defined by the recursive relations:

$$\begin{cases} A_{k0} = a_k \\ A_{kh} = \dfrac{c}{2}\left(A_{k-1,h+1} - 2\lambda A_{k-1,h} + \lambda^2 A_{k-1,h-1}\right) - \dfrac{1}{2c}\displaystyle\sum_{J=1}^{k-h+1} \psi_{J+1}\, A_{k-J,h-1} \end{cases} \tag{4.13}$$

From the integral transform theory |6|, we know that the infinite series in (4.9) can be inverted term by term in any finite interval of x, providing an entire function of exponential type. Then, by inverting (4.8), (4.9), we obtain the following representation for the Green's function:

$$G(x,t) = w_o(x)\ \delta(t - x/c) + \sum_{k=1}^{\infty} w_k(x)\ \frac{(t - x/c)^{k-1}}{(k-1)!} \tag{4.14}$$

where $\delta(\cdot)$ denotes the Dirac distribution.

In (4.14) the first term isolates the discontinuity associated with the propagating wave front, in agreement with (4.7), while the series of powers of $\tau = t - x/c$, which is uniformly convergent for any x, accounts for the response following the wave front.

We notice that, when the input function $f(t)$ is entire of exponential type, a similar wave front expansion can be carried out directly on $R(x,t)$, avoiding the convolution procedure (4.2). In this case, expanding $\bar{f}(s)$ in Laurent series

$$\bar{f}(s) = f_0/s + f_1/s^2 + \ldots \tag{4.15}$$

and performing the Cauchy product of the series (4.15), (4.3), we can write:

$$\bar{f}(s)\ a(s) = \varrho_0/s + \varrho_1/s^2 + \ldots \tag{4.16}$$

This enables us by previous considerations to obtain the series representation of $R(x,t)$:

$$R(x,t) = \sum_{k=0}^{\infty} w_k(x)\ \frac{(t - x/c)^k}{k!} \tag{4.17}$$

where the $w_k(x)$ $(k = 0,1,\ldots)$ are given by (4.12), (4.13) with $w_k(0) = A_{k0} = \varrho_k$.

5. PADE' APPROXIMANTS AND NUMERICAL CONSIDERATIONS

The series solutions (4.14), (4.17) are easy to handle for numerical computations since they are obtained in a recuresive way. A good estimate of the error, made when we truncate the series, is practically impossible, but the numerical convergence is expected to slow down by increasing of τ (time elapsed from the wave front) with a rate depend ing on x.

Since the result is an entire function in τ of exponential type which must be bounded at infinity, we are faced with the same difficulty as in the evaluation of $\exp(-\tau)$ using its Taylor expansion, when τ is large.

In his classical work |12|, Padé introduced a sequence of rational approximations, henceforth called Padé Approximants (PA), which proved to accelerate the convergence of the Taylor series for the exponential function.

No general convergence theorems are available for the PA except in the case of Stieltjes functions (see for example |13|), however the convergence in measure has been recently proved for meromorphic functions |14,15|.

In our case we apply the PA to the series solutions in τ, since the analytic properties insure their convergence in measure, and a strong improvement of the numerical convergence is expected as for the exponential function.

In the actual computations from a comparison of the partial sums of the series with the corresponding PA we have remarked that the convergence rate is much better for the PA when τ is large. Beyond a critical value $\tau=\tau_0$ the numerical convergence of the series is lost no matter how many terms are computed (we work with a fixed number of digits!) while the convergence of the PA is still satisfactory for $\tau > \tau_0$ until a matching with the long time solution is obtained.

6. RESULTS

According to the method developed in Sections 4., 5., we have performed a numerical survey of the models introduced in Sections 2., 3. The results are summarized in several tables and figures. For convenience we have fixed the time $t=T$ so that the range of the variables x, τ is finite $(0 \leq x \leq cT,\ 0 \leq \tau \leq T)$.

For the viscoelastic waves in a Standard Linear Solid (see (2.10) we have computed:

(i) the Green's function

and the solutions of the following boundary value problems:

(ii) $R_o(t) = 1$ $\qquad\qquad \bar{R}_o(s) = \frac{1}{s}$

(iii) $\qquad\qquad \bar{R}_o(s) = \frac{1}{s[1+\bar{\psi}(s)]^{\frac{1}{2}}}$

(iv) $R_o(t) = e^{-\alpha t}$ $\qquad\qquad \bar{R}_o(s) = \frac{1}{s+\alpha}$

(v) $R_o(t) = e^{-\alpha t}\cos\omega t$ $\qquad\qquad \bar{R}_o(s) = \frac{s+\alpha}{(s+\alpha)^2+\omega^2}$

In Tables I - VI we compare the series and P.A. solutions for the above boundary value problems; in order to check the method we do also quote the long time and convolution (4.2) results.

An even more stringent check for the P.A. is achieved for (iii) when $a = 0$. In this case the exact solution is explicitly known |7|:

$$R(x,t) = e^{-t/2}\, I_o\left\{\frac{1}{2}\left(t^2 - \frac{x^2}{c^2}\right)^{\frac{1}{2}}\right\}$$

and at least a five digits accuracy is found for $T \leq 50$ using no more than [12/12] P.A.

In Figures 1-4 the responses to (ii) and (iii) are plotted for several values of T.

For the thermoelastic waves the following boundary value is considered:

$$\begin{cases} \Theta_o(t) = 0 \\ E_o(t) = 1 \end{cases} \qquad\qquad \begin{cases} \bar{\Theta}_o(s) = 0 \\ \bar{E}_o(s) = \frac{1}{s} \end{cases}$$

In Tables VII, VIII we exhibit the thermal and elastic responses by comparing the series, P.A. (separately computed on the expansion of Θ^+, Θ^-, E^+, E^-) and long time |16| solutions. In Figures 5 - 8 some relevant results are shown.

From the previous examples we can infer that the Padé method plays a crucial role if we are interested in a global solution of transient wave problems.

APPENDIX

The response of a S.L.S. to any input $r_o(t)$ is given by

$$R(x,t) = \frac{1}{2\pi i}\int_{\alpha-i\infty}^{\alpha+i\infty} e^{tf(s)}\,\bar{R}_o(s)\,ds \tag{A.1}$$

where $f(s) = s\left[1 - \frac{x}{ct}\sqrt{1+\bar{\psi}(s)}\right]$, with $\bar{\psi}(s)$ given by (2.10).

For large values of t we approximate (A.1) using the saddle point method. For the Green's function $\bar{R}_o(s) = 1$ and the standard result reads

$$R(x,t) \simeq [2\pi t|f''(\bar{s})|]^{-\frac{1}{2}}\, e^{-t|f''(\bar{s})|} \tag{A.2}$$

where $\bar{s}$ is defined by $f'(\bar{s}) = 0$ and must be computed numerically.

For inputs (ii) and (iii) $\bar{R}_o(s)$ has a pole at the origin and the most relevant contribution to (A.1) is obtained when the saddle point is close to $s = 0$. Replacing $f(s)$ by $\tilde{f}(s) = f'(0)s + f''(0)s^2$ the saddle point contribution can be analytically evaluated and we get:

(ii) $R(x,t) \simeq \frac{1}{2}\,\mathrm{Erfc}[W]$ (A.3)

(iii) $R(x,t) \simeq \frac{x\sqrt{a}}{x + ct\sqrt{a}}\,\mathrm{Erfc}[W]$ (A.4)

where $W = -(1 - \frac{x}{ct\sqrt{a}}) \cdot (2\,\frac{1-a}{a} \cdot \frac{x}{ct^2\sqrt{a}})^{-\frac{1}{2}}$

For thermoelastic waves long time approximations have been derived in |16| by taking the limit as $s \to 0$ of the transform solutions and read:

$$\Theta(x,t) = -\frac{\varepsilon}{1+\varepsilon}\,\mathrm{Erf}\,[Z] \tag{A.5}$$

$$E(x,t) = 1 - \frac{\varepsilon}{1+\varepsilon}\,\mathrm{Erf}\,[Z] \tag{A.6}$$

where $Z = x\sqrt{1+\varepsilon}/2\sqrt{t}$.

REFERENCES

|1| F. MAINARDI & G. TURCHETTI, Mechanics Res. Comm., in press.

|2| F. MAINARDI & G. TURCHETTI, to be published.

|3| S. C. HUNTER, in Progress in Solid Mechanics, edited by SNEDDON & HILL, Vol. I, p. 3, North-Holland, Amsterdam, 1960.

|4| R. M. CHRISTENSEN, Theory of Viscoelasticity, Acad. Press, N.Y., 1971.

|5| J. D. ACHENBACH, "Wave Propagation in Elastic Solids", North-Holland, Amsterdam, 1973.

|6| G. DOETSCH, "Theory and Application of the Laplace Transform", Springer-Verlag, N.Y., 1974.

|7| E. M. LEE & T. KANTER, J. Appl. Phys. 24, 1115 (1956).

|8| S. KALISKY, Bull. Acad. Pol. Sci. 13, 409 (1965).

|9| E. B. POPOV, J. Appl. Math. Mech. (PMM) 31, 349 (1967).

|10) M. W. LORD & Y. SHULMAN, J. Mech. Phys. Solids 15, 299 (1967).

|11| J. D. ACHENBACH, J. Mech. Phys. Solids 16, 273 (1968).

|12| H. PADE', Thesis Ann. Ecole Nor. 9, Suppl. 1 (1892).

|13| G. A. BAKER, J. Adv. Theor. Phys. 1, 1 (1965).

|14| J. NUTTAL, J. Math. Anal. and Appl. 31, 147 (1970).

|15| J. ZINN-JUSTIN, Physics Reports (Sect. C, Phys. Lett.) 1, 55 (1971).

|16| F. R. NORWOOD & W. E. WARREN, Quart. J. Mech. Appl. Math. 22, 283 (1969).

TABLE CAPTIONS

The meaning of the symbols used in the Tables is the following:

X :	distance
TAU :	time elapsed from the wave front $\tau = t - x/c$
SERIES :	results of the partial sums (NS) of the series solution in τ .
NS :	number of terms in the partial sums
ERS :	estimated accuracy of the series solution defined by $\left\|1 - \frac{(NS-1)}{(NS)}\right\|$
PADE :	diagonal Padé approximants $[NP/NP]$ computed on the series solution in τ .
NP :	order of P.A.. Remark that $NS = 2\,NP + 1$
ERP :	estimated accuracy of P.A. defined by $\left\|1 - \frac{[NP-1/NP-1]}{[NP/NP]}\right\|$
CONVOLUTION:	convolution of the input with the Green's function computed piecewise using the series when $ERS \le 10^{-5}$, the P.A. when $ERS > 10^{-5} \ge ERP$ and the long time approximation when $ERP > 10^{-5}$. A Gauss quadrature with NG points is used.
LONG TIME :	long time approximation of the solution (see Appendix).

Viscoelastic waves in a S.L.S. with $a = .5$ for $T = 30$.

Table I	Input (i), Green's function
Table II	Input (ii)
Table III	Input (iii)
Table IV	Input (iv) with $\alpha = .1$, $NG = 8$
Table V	Input (v) with $\alpha = .1$, $\omega = \frac{\pi}{10}$, $NG = 20$

Thermal and strain waves with $\varepsilon = .03$, $\beta = 1.3$ for $T = 5$ $(NS \le 21,\ NP \le 10)$.

Table VI	Thermal waves for $E_o(t) = 1$ $\Theta_o(t) = 0$
Table VII	Elastic waves for $E_o(t) = 1$ $\Theta_o(t) = 0$

TABLE I : SLS(i) T=30

TAU	SERIES	NS	ERS	PADE'	NP	ERP	LONG TIME
0	2.593E-03	1	.0	2.593E-03	0	.0	--
2	2.123E-02	10	1.E-06	2.124E-02	4	3.E-04	2.28E-02
4	5.042E-02	12	7.E-06	5.042E-02	5	6.E-05	5.30E-02
6	7.222E-02	16	9.E-06	7.221E-02	7	3.E-06	7.53E-02
8	7.451E-02	19	1.E-05	7.451E-02	9	3.E-08	7.74E-02
10	5.973E-02	23	4.E-07	5.973E-02	11	3.E-10	6.19E-02
12	3.867E-02	25	3.E-05	3.867E-02	12	1.E-10	4.00E-02
14	2.059E-02	25	6.E-03	2.062E-02	12	1.E-08	2.13E-02
16	9.012E-03	25	2.E-02	9.124E-03	12	6.E-07	9.45E-03
18	2.021E-02	25	3.E+00	3.344E-03	12	1.E-06	3.47E-03
20	1.579E+00	25	2.E+00	1.003E-03	12	1.E-05	1.04E-03
22	3.145E+02	25	2.E+00	2.400E-04	12	3.E-04	2.50E-04
24	2.965E+04	25	2.E+00	4.911E-05	12	1.E-01	4.57E-05
26	1.516E+06	25	2.E+00	4.538E-06	12	2.E-01	5.75E-06
28	3.984E+07	25	2.E+00	2.080E-07	12	7.E-01	3.85E-07

TABLE II : SLS(ii) T = 30

TAU	SERIES	NS	ERS	PADE'	NP	ERP	LONG TIME
0	5.531E-04	1	.0	5.531E-04	0	.0	--
3	5.893E-02	11	9.E-06	5.893E-02	5	2.E-05	9.27E-02
6	2.767E-01	16	4.E-06	2.767E-01	7	5.E-07	2.49E-01
9	5.858E-01	20	8.E-06	5.858E-01	9	6.E-08	5.22E-01
12	8.288E-01	25	4.E-06	8.288E-01	12	3.E-11	8.16E-01
15	9.497E-01	25	2.E-04	9.497E-01	12	2.E-09	9.72E-01
18	1.001E+00	25	4.E-02	9.898E-01	12	5.E-09	9.99E-01
21	2.117E+01	25	2.E+00	9.986E-01	12	7.E-08	1.00E+00
24	2.788E+04	25	2.E+00	9.999E-01	12	7.E-07	1.00E+00
27	9.103E+06	25	2.E+00	1.000E+00	12	1.E-06	1.00E+00

TABLE III : SLS(iii) T = 30

TAU	SERIES	NS	ERS	PADE'	NP	ERP	LONG TIME
0	5.531E-04	1	.0	5.531E-04	0	.0	--
3	4.980E-02	12	9.E-07	4.980E-02	5	2.E-05	7.34E-02
6	2.182E-01	15	5.E-06	2.182E-01	7	3.E-08	1.87E-01
9	4.402E-01	20	8.E-06	4.402E-01	9	4.E-08	3.71E-01
12	6.031E-01	25	5.E-06	6.031E-01	12	7.E-11	5.88E-01
15	6.786E-01	25	4.E-05	6.786E-01	12	6.E-12	6.91E-01
18	7.073E-01	25	3.E-02	7.018E-01	12	8.E-09	7.07E-01
21	-3.681E+01	25	2.E+00	7.065E-01	12	2.E-09	7.07E-01
24	-9.250E+04	25	2.E+00	7.071E-01	12	2.E-06	7.07E-01
27	-6.870E+07	25	2.E+00	7.071E-01	12	3.E-06	7.07E-01

TABLE IV : SLS(iv) T = 30

TAU	SERIES	NS	ERS	PADE'	NP	ERP	CONVOLUTION
0	5.531E-04	1	.0	5.531E-04	0	.0	5.531E-04
3	5.279E-02	12	7.E-07	5.279E-02	5	5.E-05	5.279E-02
6	2.224E-01	14	9.E-06	2.224E-01	6	5.E-05	2.224E-01
9	4.054E-01	21	3.E-06	4.054E-01	10	2.E-09	4.054E-01
12	4.638E-01	25	1.E-05	4.638E-01	12	2.E-10	4.638E-01
15	3.972E-01	25	6.E-04	3.972E-01	12	1.E-08	3.972E-01
18	3.010E-01	25	2.E-01	2.867E-01	12	3.E-08	2.867E-01
21	2.240E+01	25	2.E+00	1.902E-01	12	1.E-07	1.902E-01
24	3.109E+04	25	2.E+00	1.223E-01	12	7.E-06	1.223E-01
27	1.018E+07	25	2.E+00	7.808E-02	12	9.E-06	7.808E-02

TABLE V : SLS(v) T = 30

TAU	SERIES	NS	ERS	PADE'	NP	ERP	CONVOLUTION
0	5.531E-04	1	.0	5.531E-04	0	.0	5.531E-04
3	4.835E-02	12	4.E-07	4.835E-02	5	2.E-04	4.835E-02
6	1.597E-01	16	8.E-06	1.597E-01	7	4.E-06	1.597E-01
9	1.575E-01	21	4.E-06	1.575E-01	10	5.E-08	1.575E-01
12	-1.583E-02	25	3.E-05	-1.583E-02	12	5.E-06	-1.583E-02
15	-1.400E-01	25	2.E-03	-1.399E-01	12	2.E-05	-1.399E-01
18	-6.051E-02	25	6.E-01	-7.035E-02	12	2.E-04	-7.035E-02
21	2.030E+01	25	2.E+00	5.592E-02	12	4.E-03	5.590E-02
24	2.749E+04	25	2.E+00	6.375E-02	12	5.E-03	6.556E-02
27	8.939E+06	25	2.E+00	-2.870E-02	12	1.E+00	-1.381E-02

TABLE VI : THERMAL WAVES T = 5

X	SERIES	ERS	PADE'	ERP	LONG TIME
5.24	-5.358E-02	0.	-5.358E-02	0.	--
4.81	-4.851E-02	1.E-06	-4.851E-02	3.E-06	-2.56E-02
4.18	-3.807E-02	7.E-06	-3.807E-02	1.E-05	-2.39E-02
4.18	-2.276E-02	1.E-05	-2.276E-02	2.E-05	-2.39E-02
3.54	-2.011E-02	4.E-06	-2.011E-02	8.E-06	-2.17E-02
2.90	-1.708E-02	5.E-06	-1.708E-02	3.E-07	-1.89E-02
2.27	-1.373E-02	7.E-05	-1.373E-02	6.E-07	-1.55E-02
1.63	-1.010E-02	4.E-03	-1.009E-02	9.E-06	-1.16E-02
.99	-6.707E-03	2.E-01	-6.241E-03	1.E-04	-7.30E-03
.35	-7.372E-03	4.E+00	-2.263E-03	3.E-03	-2.67E-03

TABLE VII : ELASTIC WAVES T = 5

X	SERIES	ERS	PADE'	ERP	LONG TIME
5.24	5.369E-01	0.	5.369E-01	0.	--
4.81	7.709E-01	4.E-06	7.709E-01	6.E-05	9.74E-01
4.18	9.276E-01	4.E-06	9.276E-01	1.E-05	9.76E-01
4.18	9.785E-01	4.E-06	9.785E-01	1.E-05	9.76E-01
3.54	9.811E-01	9.E-06	9.811E-01	1.E-05	9.78E-01
2.90	9.840E-01	4.E-06	9.840E-01	8.E-07	9.81E-01
2.27	9.872E-01	4.E-06	9.872E-01	7.E-09	9.84E-01
1.63	9.906E-01	3.E-04	9.906E-01	8.E-07	9.88E-01
.99	9.950E-01	1.E-02	9.942E-01	9.E-06	9.93E-01
.35	1.008E+00	3.E-01	9.979E-01	7.E-05	9.97E-01

FIGURE CAPTIONS

1 - 2 The response of a S.L.S. with $a = .5$ for input (ii) and $T = 1, 3, 5; 10, 30, 50$.

3 - 4 The same as figures 1 - 2 for input (iii)

5 - 6 Thermal and elastic waves with $\varepsilon = .03$, $\beta = 1.3$ for input $E_o(t) = 1$, $\Theta_o(t) = 0$ and $T = 2.5$.

7 - 8 The same as figures 5 - 6 for $\varepsilon = .03$, $\beta = .9$

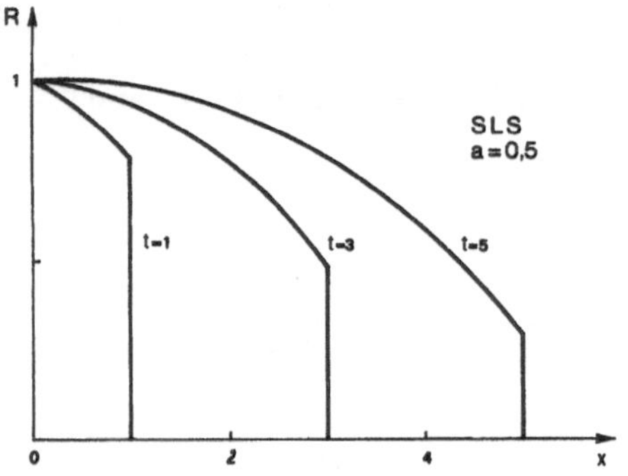

Fig. 1

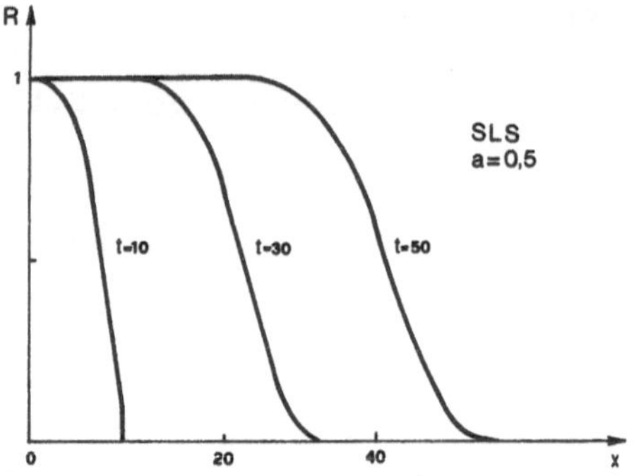

Fig. 2

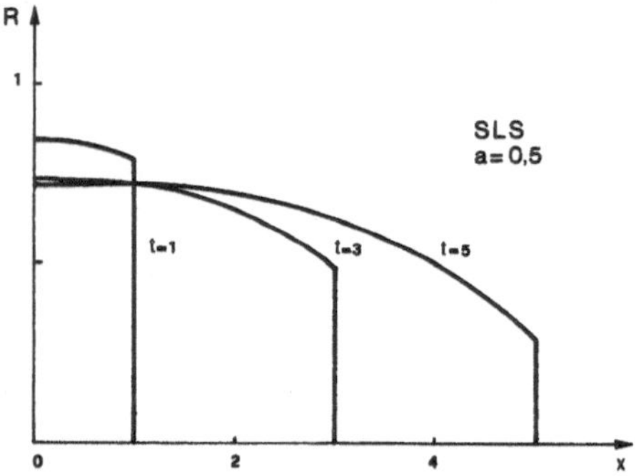

Fig. 3

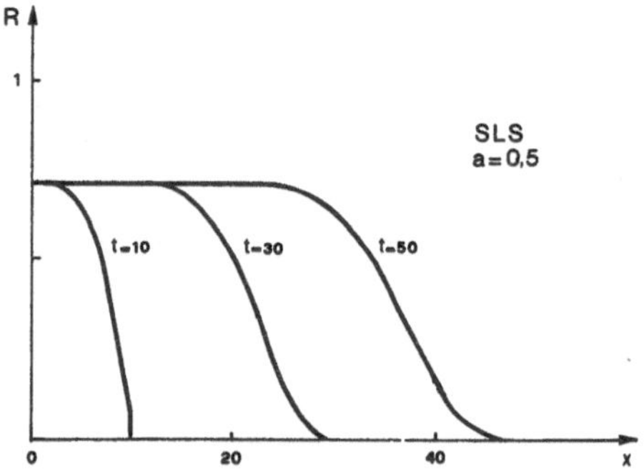

Fig. 4

Mainardi – Turchetti

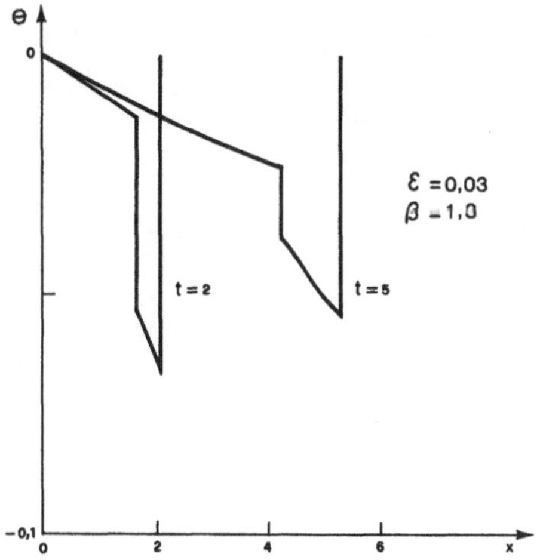

Fig. 5

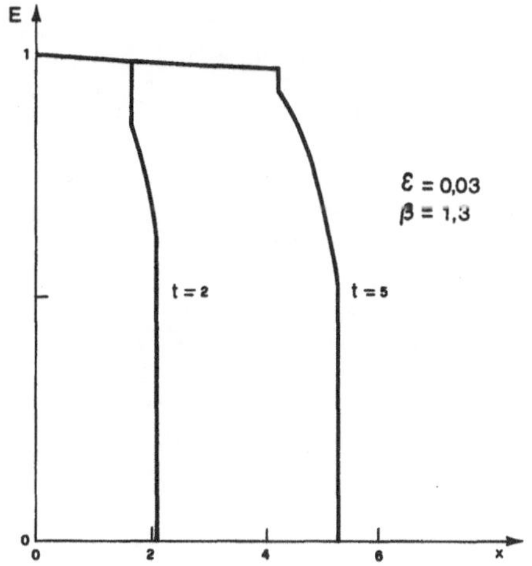

Fig. 6

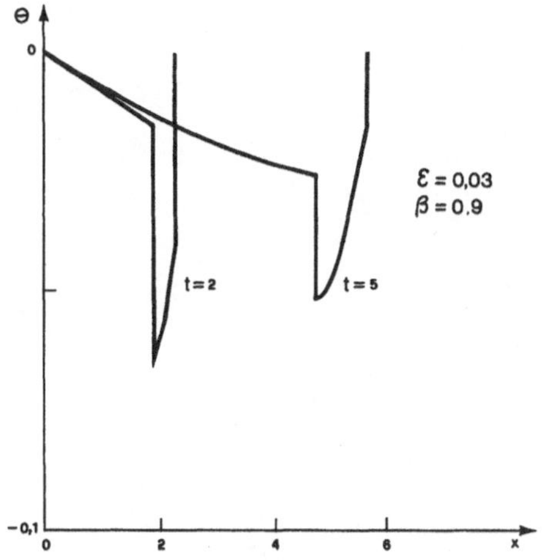

Fig. 7

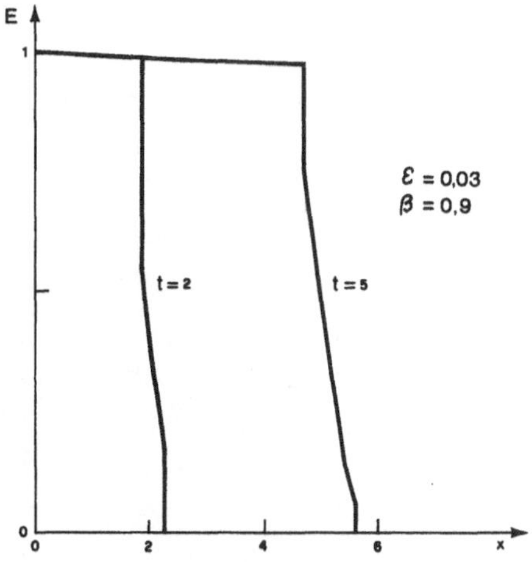

Fig. 8

APPLICATION OF METHODS FOR ACCELERATION OF CONVERGENCE TO THE CALCULATION OF SINGULARITIES OF TRANSONIC FLOWS*

Andrew H. Van Tuyl
Naval Surface Weapons Center
White Oak Laboratory
Silver Spring, Maryland 20910
USA

1. Introduction

Initial value problems in gas dynamics which lead to transonic flows include the inverse blunt body problem, in which a bow shock wave in a uniform flow is given and the body which would produce it is calculated, and the inverse calculation of nozzle flows starting from data given on the centerline. Each of these problems can be expressed as an initial value problem for a second order quasi-linear differential equation satisfied by the stream function.

When the initial curve and initial data are such that the initial curve is noncharacteristic, it follows from the Cauchy-Kowalewski theorem [1, page 39] that the initial value problem can be solved in terms of power series in the neighborhood of a given point of the initial curve. However, the region of convergence of the series obtained may be too small for practical use, due to the occurrence of singularities, either real or complex, near the initial curve. This was found by Van Dyke [2] in the case of the inverse blunt body problem, where a limiting line** (envelope of characteristics) occurs in the upstream analytic continuation of the flow. This limiting line lies closer to the shock than the distance between the latter and the body, and hence, a power series solution in the neighborhood of a point of the shock diverges at the body and cannot be used directly to calculate the flow there.

In [3], Leavitt has calculated the shape and position of this limiting line near the axis of symmetry by a modification of a method due to Domb [4], starting

* This work was supported by the Naval Surface Weapons Center Independent Research Fund.
**Also called limit line.

from the power solution of the inverse blunt body problem in the neighborhood of the nose of the shock. The location of the limiting line was then used to transform the series so that convergence was obtained at the body. More recently, Schwartz [5] has used Domb's method to calculate limiting lines in the flows produced by parabolic and paraboloidal shocks in a free stream of infinite Mach number. Various modifications and extensions of Domb's method have been applied to problems of statistical mechanics by Domb, Sykes, Fisher, and others ([6], for example).

Limiting lines in solutions of the inverse blunt body problem have also been calculated by Garabedian and his students ([7] and [8]) by use of Garabedian's method of complex characteristics.

Limiting lines may also occur in nozzle flows obtained from given centerline distributions of velocity or Mach number. As in the case of blunt body flows, the region of convergence of a power series solution may be restricted by a limiting line even though the point about which the solution is obtained lies in the subsonic region. In nozzle design, it is of practical interest to know if a given centerline distribution leads to a limiting line which lies between a desired streamline and the axis of symmetry.

A procedure for calculation of limiting lines will be described, starting from a power series solution, in which methods for acceleration of convergence are used. This procedure involves the ratio of successive coefficients of a power series, as in Domb's method, and a necessary requirement is therefore that the extent of the region of convergence in the direction of at least one of the coordinate axes should be determined by a limiting line. Sequences are constructed which converge to points on a limiting line and to its order $k \geq 1$. With the assumption that the single power series used in this calculation has only one singularity on its circle of convergence, it is proved that certain nonlinear sequence transformations, including the $e_1^{(s)}$ transformation defined by Shanks ([9], page 39) accelerate the convergence of these sequences.

The results obtained hold also for analytic initial value problems for other equations or systems of equations in two independent variables, when the given equation or system of equations can be replaced by a characteristic system in two independent variables. In particular, limiting lines can be calculated by the present method in the one-dimensional unsteady flow produced by a given piston motion. Finally, the present method is also applicable to some of the series occurring in [6].

2. Limiting Lines of Order k

In both the inverse blunt body problem and the inverse calculation of nozzle flows, the stream function ψ satisfies a quasi-linear second order partial differential equation of the form

$$a\psi_{xx} + b\psi_{xy} + c\psi_{yy} + d = 0, \tag{2.1}$$

where the coefficients are analytic functions of their arguments. The independent variables denote cartesian coordinates in the two-dimensional case and cylindrical coordinates in the axially symmetric case. As in [1], pp. 491-493, (2.1) can be replaced by the system of characteristic equations

$$\begin{aligned} y_\alpha &= h_1 x_\alpha \\ y_\beta &= h_2 x_\beta \\ p_\alpha + h_2 q_\alpha + \frac{d}{a} x_\alpha &= 0 \\ p_\beta + h_1 q_\beta + \frac{d}{a} x_\beta &= 0 \\ \psi_\alpha - p x_\alpha - q y_\alpha &= 0 \end{aligned} \tag{2.2}$$

where h_1 and h_2 are the roots of the equation

$$ah^2 - bh + c = 0. \tag{2.3}$$

Real values of α and β correspond to values of x and y for which (2.1) is hyperbolic. It follows from the Cauchy-Kowalewski theorem that the solution of an analytic initial value problem for (2.2) in the real $\alpha\beta$-plane is analytic.

Given a solution of (2.2) which is analytic in a domain D of the real $\alpha\beta$-plane, the functions $x(\alpha,\beta)$ and $y(\alpha,\beta)$ define a mapping which is one-to-one in any portion of D in which the Jacobian $J = x_\alpha y_\beta - x_\beta y_\alpha$ does not vanish. Let $k \geq 1$ be an integer, and let J and its derivatives of order up to and including k-1 vanish along a curve C in D. Then the image of C in the xy-plane is defined to be a limiting line of order k.

The well-known result ([10], for example) that a regular arc of a limiting line of the first order is an envelope of one of the families of characteristics can also be shown to hold for limiting lines of order $k > 1$. As in the case of limiting lines of first order, characteristics of the second family have infinite curvature at the limiting line for $k>1$.

Finally, we can prove also that the behavior of flow quantities in the neighborhood of a limiting line of order $k \geq 1$ is given by the following theorem:

Theorem 1. Let a solution of (2.1) have a limiting line of order $k \geq 1$ with the equation $x = x_0(y)$, where $x_0(y)$ is analytic for $y_1<y<y_2$, and let the solution be analytic for $x_1<x<x_0(y)$, $y_1<y<y_2$. Let $F(x,y)$ be any one of the flow variables. Then $F(x,y)$ has an expansion of the form

$$F(x,y) = \sum_{n=0}^{\infty} a_n(y) \left[1 - \frac{x}{x_0(y)}\right]^{n/(k+1)}$$

for $y_1<y<y_2$ and for $x_0(y)-x$ sufficiently small.

3. An Asymptotic Result

An asymptotic result on which the present method of calculation is based is given by the following theorem:

Theorem 2. Let f(z) be analytic for $|z| \leq 1$ except at $z = 1$, with

$$f(z) = \sum_{n=0}^{\infty} a_n z^n, \quad |z| < 1$$

$$= \sum_{n=0}^{\infty} b_n (1-z)^{n/m}, \quad |1-z| < \eta, \quad |\arg(1-z)| < \pi,$$

where $m=2,3,\ldots$ is an integer. Then as $n \to \infty$, we have

$$a_n = \binom{1/m}{n} \left\{ b_0 + \sum_{j=1}^{N-1} c_j n^{-j/m} + 0(n^{-N/m}) \right\}$$

for any $N \geq 2$, where $c_j = 0$ when $j + 1$ is divisible by m. When $N + 1$ is divisible by m, the error is $0(n^{-(N+1)/m})$.

Proof. Let

$$f(z) = \sum_{j=0}^{N+m-1} b_j (1-z)^{j/m} + f_1(z).$$

Then $f_1(z)$ is analytic for $|z| \leq 1$ except at $z = 1$. Writing $r=[N/m]+1$, we see that derivatives of $f_1(z)$ up to and including the rth are continuous on $|z|=1$, while the (r+1)st is discontinuous on $|z|=1$ at $z=1$, but integrable there. Then with

$$f_1(z) = \sum_{n=0}^{\infty} a_n^{(1)} z^n, \quad |z| < 1,$$

it is easily shown by use of Cauchy's theorem that $|a_n^{(1)}|=0(n^{-r-1})$ as $n \to \infty$. The stronger result $|a_n^{(1)}|=o(n^{-r-1})$ can be shown to hold by use of the Riemann-Lebesgue lemma, but the weaker result is sufficient for the present purpose.

Let N be such that $b_N \neq 0$. Then

$$a_n = \sum_{j=0}^{N-1} b_j \binom{j/m}{n} + R_{N,n},$$

where

$$R_{N,n} = \sum_{j=N}^{N+m-1} b_j \binom{j/m}{n} + a_n^{(1)}.$$

It follows from Stirling's approximation that

$$\lim_{n\to\infty} n^{\lambda+1}\binom{\lambda}{n} < \infty$$

for $\lambda > o$. Hence, since $[N/m] + 1 > N/m$, we have

$$\lim_{n\to\infty} a_n^{(1)}/\binom{N/m}{n} = 0.$$

It follows that when $b_N \neq o$, $R_{n,N} = O(n^{-N/m-1})$. Finally, by use of Stirling's approximation, we obtain the result stated.

Corollary 1. When $b_o \neq o$, we have

$$a_{n+1}/a_n = \left(1 - \frac{3/2}{n+1}\right)\left\{1 + \sum_{j=1}^{N-1} d_j n^{-j-1} + O(n^{-N-1})\right\}, \quad m = 2$$

$$= \left(1 - \frac{1+1/m}{n+1}\right)\left\{1 + \sum_{j=1}^{N-1} d_j n^{-j/m-1} + O(n^{-N/m-1})\right\}, \quad m > 2.$$

Corollary 2. When $b_o = 0$ but $b_1 \neq o$,

$$a_{n+1}/a_n = \left(1 - \frac{5/2}{n+1}\right)\left\{1 + \sum_{j=1}^{N-1} e_j n^{-j-1} + O(n^{-N-1})\right\}, \quad m = 2$$

$$= \left(1 - \frac{1+2/m}{n+1}\right)\left\{1 + \sum_{j=1}^{N-1} e_j n^{-j/m-1} + O(n^{-N/m-1})\right\}, \quad m > 2.$$

A similar but more complicated theorem can be proved when there are several singularities on $|z|=1$, with a different value of m associated with each. With more than one singularity, however, the transformations used here become less effective.

4. Sequences for Calculation of Limiting Lines

Given an analytic initial value problem for an equation of the form of (2.1), let the origin be taken at a point of the initial curve, and let $(x_o, 0)$ be a point on a limiting line of order $k \geq 1$. Let $F(x,y)$ be any one of the dependent variables, such as the density or pressure in the inverse blunt body problem, and as described in the introduction, let $F(z,0)$ be analytic for $|z| \leq x_o$ except at $z = x_o$. This assumption, that only one singularity lies on the circle of convergence, appears to be verified in special cases which have been calculated.

We have

$$F(z,o) = \sum_{n=0}^{\infty} a_n z^n, \quad |z| < x_o, \tag{4.1}$$

and by Theorem 1,

$$F(z,o) = \sum_{n=0}^{\infty} b_n \left(1 - \frac{z}{x_o}\right)^{n/(k+1)} \tag{4.2}$$

in a cut neighborhood of $z = x_o$. It follows from Corollary 1 of Theorem 2 that when $b_o \neq 0$ in (4.2),

$$\begin{aligned}\left(1 - \frac{1+1/(k+1)}{n+1}\right)\frac{a_n}{a_{n+1}} &= x_o \left\{1 + \sum_{j=1}^{N-1} c_j n^{-j-1} + 0(n^{-N-1})\right\}, \quad k = 1 \\ &= x_o \left\{1 + \sum_{j=1}^{N-1} c_j n^{-j/(k+1)-1} + 0(n^{-N/(k+1)-1})\right\}, \quad k > 2\end{aligned} \tag{4.3}$$

From (4.3), we obtain

$$\begin{aligned} r_n = (s_n-1)(n+1)(n+2) &= \frac{3}{2}\left\{1 + \sum_{j=1}^{N-1} d_j n^{-j} + 0(n^{-N})\right\}, \quad k = 1 \\ &= (1+1/(k+1))\left\{1 + \sum_{j=1}^{N-1} d_j n^{-j/(k+1)} + 0(n^{-N/(k+1)})\right\}, \quad k \geq 2 \end{aligned} \tag{4.4}$$

when $b_o \neq 0$, where $s_n = a_{n+2}a_n/a_{n+1}^2$. Noting that $(n+2)s_n-(n+1)$ tends to unity as $n \to \infty$, we see that the sequence

$$t_n = \frac{(s_n-1)(n+1)(n+2)}{(n+2)s_n-(n+1)} \tag{4.5}$$

tends to the same limit as the left hand side of (4.4), and we can show that it has the same asymptotic form as (4.4) as $n \to \infty$. Denoting the coefficients of the asymptotic expression for t_n by e_j, we find that $e_1 = d_1$ when $k \geq 2$. When $k = 1$, however, we have $d_1 = e_1 + 3/2$. Thus, while t_n and r_n have the same rate of convergence, one may be asymptotically more accurate than the other.

When $b_o=0$ but $b_1\neq 0$, sequences similar to the preceding can be obtained by use of Corollary 2 of Theorem 2. By consideration of both the limit and the rate of convergence of t_n, it is possible to determine k and to decide whether or not b_o vanishes.

5. Nonlinear Sequence Transformations

Let A_n be a given sequence, and let

$$B_n = \frac{A_{n+1}A_{n-1}-A_n^2}{A_{n+1}+A_{n-1}-2A_n}. \tag{5.1}$$

The sequence B_n is equal to the first order transform $e_1(A_n)$ defined by Shanks [9] and to the sequence $\varepsilon_2^{(n)}$ obtained by the ε-algorithm of Wynn [11], and is also referred to as the Aitken δ^2 process. Under certain conditions, a convergent sequence A_n is transformed to a sequence B_n which has the same limit as A_n and converges more rapidly. In particular, the latter holds when $\lim_{n\to\infty} |\Delta A_{n+1}/\Delta A_n|\neq 1$. When $\lim_{n\to\infty} |\Delta A_{n+1}/\Delta A_n|=1$ but $\lim_{n\to\infty} \Delta A_n/\Delta B_n=s\neq 1$, the transformed sequence B_n converges with the same rapidity as A_n. For this case, Shanks ([9], page 39) has defined the transform

$$e_1^{(s)}(A_n) = \frac{sB_n-A_n}{s-1}. \tag{5.2}$$

A transformation equivalent to this, called U_n, was also introduced by Lubkin [12].

Finally, Lubkin ([12], page 229) has introduced a more general transformation which is expressed in the present notation by

$$W_n = \frac{B_n\dfrac{\Delta A_n}{\Delta B_n} - A_n}{\dfrac{\Delta A_n}{\Delta B_n} - 1} \tag{5.3}$$

With the preceding definitions, we can verify the following theorems by direct calculation:

<u>Theorem 3</u>. Let a given sequence have the form

$$A_n = A + \sum_{j=0}^{N-1} a_j n^{-j-\alpha} + O(n^{-N-\alpha})$$

as $n\to\infty$ for all $N \geq 1$. Then

$$e_1^{(\alpha+1)}(A_n) = A + \sum_{j=0}^{N-1} b_j n^{-j-\alpha-2} + O(n^{-N-\alpha-2}).$$

<u>Corollary</u>. With the notation $e_1^{(s_1)}(e_1^{(s_2)}(A_n)) = e^{(s_1)}e_1^{(s_2)}(A_n)$, we have

$$e^{(\alpha+2r+1)}e_1^{(\alpha+2r-1)}\cdots e_1^{(\alpha+1)}(A_n) = A + O(n^{-\alpha-2r-2}).$$

<u>Theorem 4</u>. Let a given sequence have the form

$$A_n = A + \sum_{j=1}^{N-1} a_j n^{-j/m-\alpha} + O(n^{-N/m-\alpha})$$

as $n\to\infty$ for all $N \geq 2$ where $m = 2,3,\ldots$. Then

$$e_1^{(1+\alpha+1/m)}(A_n) = A + \sum_{j=2}^{N-1} b_j n^{-j/m-\alpha} + O(n^{-N/m-\alpha})$$

for $N = 3,4,\ldots$.

<u>Corollary</u>.

$$e_1^{(1+\alpha+r/m)}e_1^{(1+\alpha+(r-1))/m}\cdots e_1^{(1+\alpha+1/m)}(A_n) = A + O(n^{-(r+1)/m-\alpha}).$$

Theorems 3 and 4 have direct application to the sequences of Section 4. In particular, with

$$A_n = (1 - \frac{3/2}{n+1})\frac{a_n}{a_{n+1}}$$

in the case k = 1, we see that

$$e_1^{(2r+1)} e_1^{(2r-1)} \ldots e_1^{(3)}(A_n) = x_o + O(n^{-2r-2}).$$

We see that the sequences A_n in Theorems 3 and 4 must have $\alpha > 0$ and $\alpha > -1/m$, respectively, in order to be convergent. However, the theorems do not specify the sign of α, and the corollaries show that the iterated transformations converge to A for sufficiently large values of r whether or not A_n converges.

Finally, the following theorems for W_n are nearly identical with the preceding:

Theorem 5. Let a given sequence have the form

$$A_n = A + \sum_{j=o}^{N-1} a_j n^{-j-\alpha} + O(n^{-N-\alpha})$$

as $n\to\infty$ for N = 1,2,... . Then

$$W_n = A + \sum_{j=o}^{N-1} c_j n^{-j-\alpha-2} + O(n^{-N-\alpha-2}),$$

Theorem 6. Let a given sequence have the form

$$A_n = A + \sum_{j=1}^{N-1} a_j n^{-j/m-\alpha} + O(n^{-N/m-\alpha}).$$

as $n\to\infty$ for N = 2,3,... where m = 2,3,... . Then

$$W_n = A + \sum_{j=2}^{N-1} c_j n^{-j/m-\alpha} + O(n^{-N/m-\alpha}).$$

for N = 3,4... . Corollaries which are exactly parallel to the corollaries of Theorems 3 and 4 clearly hold.

6. Numerical Examples

The sequences of Section 4, together with the sequence transformations of Section 5, have been applied to the calculation of limiting lines of first order in the inverse blunt body problem and in the calculation of nozzle flows. The calculation of the axial point on the upstream limiting line in the case of a paraboloidal shock at a free stream Mach number of 2 is shown in Table 1, which was calculated by use of 39 coefficients of the power series for the density. Calculations were carried out on the CDC 6500 in double precision, using the method of [13]. The sequences r_n and t_n converge to 3/2 and are accelerated by the transformation $e_1^{(2)}$, showing that the upstream limiting line is of first order, and that $b_o \neq 0$ in the expansion of the density in the neighborhood of the limiting line. Values of r_n, t_n, and $e_1^{(2)}(t_n)$ are given in Table 2, and show that t_n is preferable to r_n in this case. When the same calculations are repeated using the power series for the stream function, it is found that r_n and t_n approach the limit 5/2. Thus, it appears that $b_o=0$ while $b_1 \neq 0$ in the expansion of the stream function in the neighborhood of the axial point of the limiting line.

The rates of convergence shown in Section 5 cannot usually be realized beyond the transformation $e_1^{(3)}$ when using single precision, because of loss of significance due to subtraction. However, the accuracy obtained by use of $e_1^{(3)}$ in single precision has been found to be sufficient in all examples calculated. Agreement of successive terms of the sequence to 5 or 6 significant figures is usually obtained.

Finally, calculations of limiting lines in nozzle flows have been carried out in single precision on the CDC 6500 using $e_1^{(3)}$. The result of such a calculation is shown in Figure 1 in the axially symmetric case, where $u = 1 + (\sqrt{3}/2)x$ on the centerline. This centerline velocity distribution is the same as that of case (f) of [14] after a change of reference quantities. The sonic line, limiting characteristic, and streamlines were calculated by the method of [15]. All calculations were carried out using 25 terms of the series in y, and convergence to 5 or more figures was found except near the limiting line. With the use of double precision, the more rapidly convergent transformation $e_1^{(5)} e_1^{(3)} (A_n)$ would give additional accuracy.

The portion of the streamline $\psi = 0.2$ to the left of the limiting characteristic can be taken as the wall of a nozzle in the subsonic and transonic regions, since the flow is analytic on the limiting characteristic up to and on the streamline. However, the streamline $\psi = 0.45$ meets the limiting line above its point of tangency with the limiting characteristic, and cannot be used as part of a nozzle contour. The dashed portion of the limiting characteristic belongs to a second solution of the flow equations having the limiting line shown. The two solutions correspond to the same analytic solution of equations (2.2) in the $\alpha\beta$-plane, but on opposite sides of the curve $J = 0$ which corresponds to the limiting line.

7. References

1. R. Courant and D. Hilbert, Methods of Mathematical Physics, Vol. II, Interscience Publishers, New York, 1962.

2. M. D. Van Dyke, "A Model of Supersonic Flow Past Blunt Axisymmetric Bodies with Application to Chester's Solution," Journal of Fluid Mechanics, Vol. 3, (1958), pp. 515-522.

3. J. A. Leavitt, "Computational Aspects of the Detached Shock Problem," AIAA Journal, Vol. 6 (1968), pp. 1084-1088.

4. C. Domb, "Order-Disorder Statistics, II. A Two Dimensional Model," Proceedings of the Royal Society of London, Series A, Vol. 199 (1949), pp. 199-221.

5. L. W. Schwartz, "Hypersonic Flows Generated by Parabolic and Paraboloidal Shock Waves," Physics of Fluids, Vol. 17 (1974), pp. 1816-1821.

6. M. E. Fisher, "The Nature of Critical Points," Lectures in Theoretical Physics, Vol. VIIC, University of Colorado Press, Boulder, 1964, pp. 1-159.

7. P. R. Garabedian, "Global Structure of Solutions of the Inverse Detached Shock Problem," Problems of Hydrodynamics and Continuum Mechanics, published by the Society for Industrial and Applied Mathematics, 1969, pp. 318-322.

8. E. V. Swenson, "Geometry of the Complex Characteristics in Transonic Flow," Comm. Pure and App. Math., Vol. 21 (1968), pp. 175-185.

9. D. Shanks, "Non-Linear Transformations of Divergent and Slowly Convergent Sequences," Journal of Mathematics and Physics, Vol. 34 (1955), pp. 1-42.

10. R. Courant and K. O. Friedrichs, Supersonic Flow and Shock Waves, Interscience Publishers, New York, 1948.

11. P. Wynn, "On a Device for Computing the $e_m(S_n)$ Transform," MTAC, Vol. 10 (1956), pp. 91-96.

12. S. Lubkin, "A Method of Summing Infinite Series," Journal of Research of the Bureau of Standards, Vol. 48 (1952), pp. 228-254.

13. A. H. Van Tuyl, "Use of Padé Fractions in the Calculation of Blunt Body Flows," AIAA Journal, Vol. 9 (1971), pp. 1431-1433.

14. G. V. R. Rao and B. Jaffe, "A Numerical Solution of Transonic Flow in a Convergent-Divergent Nozzle," Final Rept., Contract NAS7-635, March 1969, NASA.

15. A. H. Van Tuyl, "Calculation of Nozzle Flows Using Padé Fractions," AIAA Journal, Vol. 11 (1973), pp. 537-541.

Table 1. Distance of Upstream Limiting Line from Nose of Shock for Paraboloidal Shock Wave $r^2 = 2x$ with Free Stream Mach Number = 2.

n	$A_n = -(1 - \frac{3/2}{n+1})\frac{a_n}{a_{n+1}}$	$e_1^{(3)}(A_n)$	$e_1^{(5)}e_1^{(3)}(A_n)$
34	0.083546753454	0.0835451981166	0.0835451972856
35	.0835466622853	.0835451980201	.0835451972406
36	.0835465789026	.0835451979365	.0835451972874
37	.0835465024447	.0835451978019	.0835451972615

Table 2. Sequences r_n and t_n Corresponding to Axial Point on Upstream Limiting Line for Paraboloidal Shock Wave $r^2 = 2x$ with Free Stream Mach Number = 2.

n	r_n	t_n	$e_1^{(2)}(t_n)$
34	1.5685409	1.5012612	1.5000020
35	1.5665483	1.5012223	1.5000018
36	1.5646684	1.5011858	1.5000017
37	1.5628918	1.5011514	1.5000016

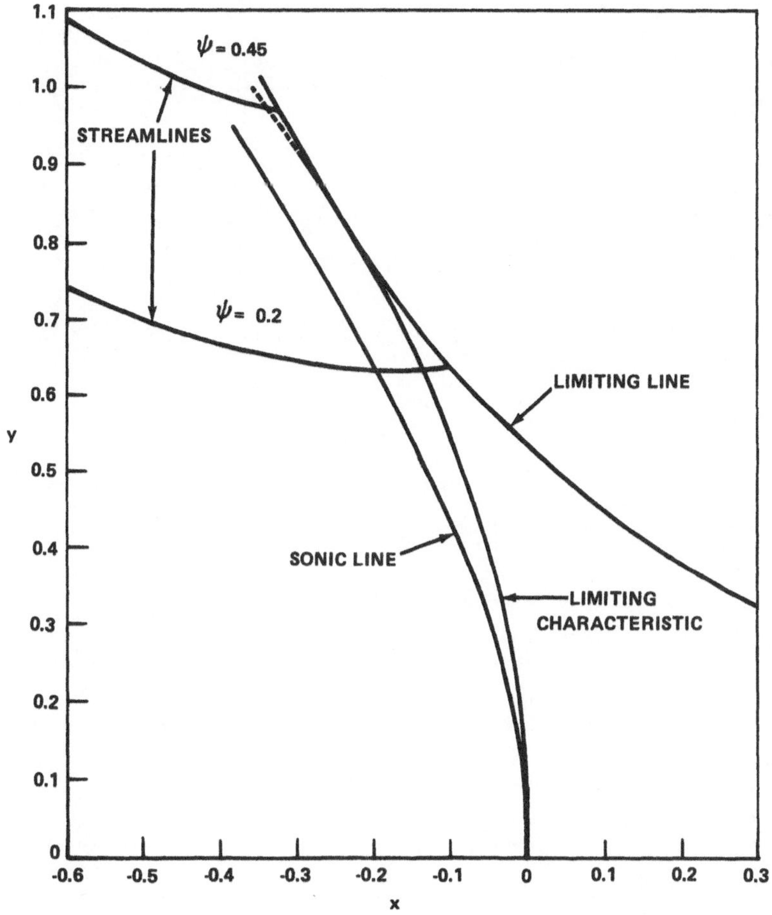

Figure 1. Limiting line in the nozzle flow with centerline velocity distribution $u = 1 + (\sqrt{3}/2)x$.

THE USE OF PADE FRACTIONS IN THE CALCULATION OF NOZZLE FLOWS*

Andrew H. Van Tuyl
Naval Surface Weapons Center
White Oak Laboratory
Silver Spring, Maryland 20910
USA

1. Introduction

In nozzle design, flow calculations are usually carried out by an inverse procedure in which either the Mach number or velocity is given on the centerline. Calculations in the supersonic region can be carried out accurately by the method of characteristics when the solution in the transonic region is known. However, the inverse problem is improperly-posed in the subsonic region, and it is therefore not possible to calculate the subsonic and transonic portions of the flow to arbitrary accuracy by means of finite-difference marching procedures without the use of complex extension. Methods for calculating the flow in the transonic region ([1], for example) are mainly applicable to the design of nozzles with large throat radius of curvature, such as wind tunnel nozzles. These methods are not sufficiently accurate for the design of short nozzles, since rapid changes then occur near the throat. Improved accuracy in the case of short nozzles is given by [2], but the accuracy which can be obtained is limited by the fixed number of terms of the power series used. Short nozzles with a uniform exit flow are of interest in connection with gas dynamic lasers [3], where two-dimensional nozzles have been used, and with chemical lasers [4], where both two-dimensional and axially symmetric nozzles are applicable. More accurate subsonic calculations have been carried out in the axially symmetric case by Armitage [5] and Rao and Jaffe [6] by means of Garabedian's method of complex characteristics [7, 8]. Calculations by the method of complex characteristics have also been carried out by Solomon [9] in the two-dimensional and axially symmetric cases.

*This work was supported by the Naval Surface Weapons Center Independent Research Fund.

When the flow quantity given on the centerline is analytic, it follows from the Cauchy-Kowalewski theorem ([10], page 39) that the inverse problem can be solved in terms of power series in the neighborhood of a given point of the centerline. While these series are known to converge near the given point, the region of convergence may be such that the series are either divergent or too slowly convergent at points of physical interest. It was noted by Van Dyke [11] that the former occurs in the case of the inverse blunt body problem, due to the presence of a limiting line (envelope of characteristics) in the upstream analytic continuation of the flow. This limiting line lies closer to the shock than the distance between the shock and the body, and hence, the body does not lie within the region of convergence of the series. However, it was found in [12], [13], and [14] that Padé fractions formed from these series give accurate results at the body, and can be used to compute the flow there. A similar procedure has been used in [15], where Padé fractions are formed from certain power series in the neighborhood of points of the centerline. As in the case of the inverse blunt body problem, examples indicate that the use of Padé fractions leads to convergence when the series diverge, and that convergence is accelerated when the latter converge. The region of convergence of the series may be restricted by limiting lines, as in the blunt body problem, or by complex singularities. Accurate calculations can also be carried out in the supersonic region by use of Padé fractions when limiting lines do not occur near the centerline. Near a limiting line, convergence of both the series and the sequences of Padé fractions is found to be slow.

The present paper extends [15] by investigating additional ways in which Padé fractions can be used in the calculation of nozzle flows. In particular, Padé fractions are formed from power series along given curves which intersect the centerline. By use of these results, more efficient use of the double Taylor expansions obtained in [15] can be made.

2. Power Series Solution of the Inverse Problem

The coordinate system used is shown in Figure 1, where x and y denote Cartesian coordinates in the two-dimensional case and cylindrical coordinates in the axially symmetric case, and where the origin is at the sonic point on the centerline. We will assume irrotational flow of a perfect gas with ratio of specific heats γ. Let $\rho^* = c^* = 1$, where ρ^* is the critical density and c^* is the critical sound speed, and let the stream function satisfy

$$u = (1/\rho y^\sigma)\partial\psi/\partial y, \quad v = -(1/\rho y^\sigma)\partial\psi/\partial x, \tag{2.1}$$

where u and v are the x and y components of velocity, respectively, ρ is the density, and $\sigma = 0$ and 1, respectively, in the two-dimensional and axially symmetric cases. Substituting (2.1) into Bernoulli's equation and the condition for irrotationality, we obtain

$$(1/y^{2\sigma})(\psi_x{}^2 + \psi_y{}^2) + [2/(\gamma-1)]\rho^{\gamma+1} - [(\gamma+1)/(\gamma-1)]\rho^2 = 0 \tag{2.2}$$

and

$$\rho(\psi_{xx} + \psi_{yy} - \sigma y^{-1}\psi_y) - \rho_x\psi_x - \rho_y\psi_y = 0, \tag{2.3}$$

respectively.

Let the density, Mach number, and velocity on the centerline be denoted by $\rho_o(x)$, $M_o(x)$, and $u_o(x)$, respectively. Then $\rho_o(x)$ is given in terms of $M_o(x)$ by the relation

$$\rho_o = \left\{2/(\gamma+1) + [(\gamma-1)/(\gamma+1)]M_o{}^2\right\}^{-1/(\gamma-1)}, \tag{2.4}$$

and in terms of $u_o(x)$, by

$$\rho_o = \left\{(\gamma+1)/2 - [(\gamma-1)/2]u_o{}^2\right\}^{1/(\gamma-1)}. \tag{2.5}$$

Let $M_o(x)$ and $u_o(x)$ have the expansions

$$M_o(x) = \sum_{i=1}^{\infty} M_i(x-x_1)^{i-1} \tag{2.6}$$

and

$$u_o(x) \sum_{i=1}^{\infty} u_i(x-x_1)^{i-1} \tag{2.7}$$

in the neighborhood of $x = x_1$. Given either (2.6) or (2.7), we can find the coefficients ρ_i of the expansion

$$\rho_o(x) = \sum_{i=1}^{\infty} \rho_i (x-x_1)^{i-1} \tag{2.8}$$

by use of (2.4) or (2.5) respectively, and subroutines for power series manipulations. We can then find the solution of the inverse problem in the neighborhood of the point $(x_1, 0)$ in the form

$$\psi = \sum_{i=1}^{\infty} \sum_{j=1}^{\infty} \psi_{ij}\, y^{2i-1+\sigma} (x-x_1)^{j-1}, \tag{2.9}$$

$$\rho = \sum_{i=1}^{\infty} \sum_{j=1}^{\infty} \rho_{ij}\, y^{2i-2} (x-x_1)^{j-1}. \tag{2.10}$$

We see that $\rho_{1j} = \rho_j$. To summarize the calculation of the remaining coefficients, let the left hand sides of (2.2) and (2.3) be denoted by A and B, respectively. We have

$$A = \sum_{i=1}^{\infty} \sum_{j=1}^{\infty} a_{ij}\, y^{2i-2} (x-x_1)^{j-1} \tag{2.11}$$

and

$$B = \sum_{i=1}^{\infty} \sum_{j=1}^{\infty} b_{ij}\, y^{2i-1+\sigma} (x-x_i)^{j-1} \tag{2.12}$$

For $j \geq 2$, we find that

$$a_{1j} = 2(1+\sigma)^2 \psi_{11} \psi_{1j} + a'_{1j}, \tag{2.13}$$

where

$$\psi_{11} = [1/(1+\sigma)] \left\{ [(\gamma+1)/(\gamma-1)] \rho_{11}{}^2 - [2/(\gamma-1)] \rho_{11}{}^{\gamma+1} \right\}^{1/2} \tag{2.14}$$

and where a'_{1j} does not involve ψ_{1j}. For $i \geq 2$, we have

$$\begin{aligned} a_{ij} &= C_1\, \psi_{ij} + C_2\, \rho_{ij} + a'_{ij}, \\ b_{ij} &= C_3\, \psi_{ij} + C_4\, \rho_{ij} + b'_{ij}, \end{aligned} \tag{2.15}$$

where

$$
\begin{aligned}
C_1 &= 2(1+\sigma)(2i-1+\sigma)\psi_{11}, \\
C_2 &= [2(\gamma+1)/(\gamma-1)](\rho_{11}{}^{\gamma}-\rho_{11}), \\
C_3 &= 2(2i-1+\sigma)(i-1)\rho_{11}, \\
C_4 &= -2(1+\sigma)(i-1)\psi_{11}.
\end{aligned} \tag{2.16}
$$

and where a'_{ij} and b'_{ij} do not involve ψ_{ij} and ρ_{ij}. The calculation of the remaining coefficients can now be described as follows: First, ψ_{1j} is calculated by setting $a_{1j} = 0$ in (2.13) for $j = 2, 3, \ldots$. For each j, a'_{1j} is found by setting $\psi_{1j} = 0$ and calculating a_{1j}. The coefficients ψ_{ij} and ρ_{ij} are then found for $i \geq 2$ by setting a_{ij} and b_{ij} equal to zero in (2.15) and solving the resulting pair of equations simultaneously. For given i and j, a'_{ij} and b'_{ij} are obtained by setting ψ_{ij} and ρ_{ij} equal to zero and calculating a_{ij} and b_{ij}. Assuming that ρ_i is known for $1 \leq i \leq 2K - 1$, we can find ψ_{ij} and ρ_{ij} for $1 \leq i \leq K$ and $1 \leq j \leq 2K - 2i + 1$.

3. Pade´ Fractions

Let

$$f(z) = \sum_{i=0}^{\infty} c_i z^i \tag{3.1}$$

be a given power series with $c_o \neq 0$. Then Pade´ fractions $f_{k,n}(z)$, $k \geq 0$, $n \geq 0$, are defined as follows: $f_{k,n}(z)$ is a rational fraction with numerator and denominator of degrees less than or equal to n and k, respectively, such that the Taylor expansion of $f_{k,n}(z)$ in the neighborhood of the origin agrees with (3.1) to more terms than that of any other rational fraction with numerator of degree $\leq n$ and denominator of degree $\leq k$ ([16], page 377). This fraction always exists and is unique. Convenient methods of calculation include the QD algorithm of Rutishauser [17] and the ε - algorithm of Wynn [18].

The sequence $f_{n,n}(z)$, which involves the first $2n + 1$ terms of (4.1), is often found to converge much more rapidly than sequences in which either k or n remains constant. The convergence of $f_{n,n}(z)$ has been proved only for special classes of functions, and none of the available convergence results appears to be directly applicable to the series occurring in either the inverse blunt body problem or the inverse calculation of nozzle flows. In both [14] and [15], however, it was found that the sequence $f_{n,n}(z)$ converges in most cases when the series diverges, and that it accelerates convergence when the latter converges.

4. Power Series for Constant x

Before Padé fractions can be used, the solution of the inverse problem must be expressed in terms of power series in one variable. One approach is to write (2.9) and (2.10) as single power series in y with coefficients which are functions of x. We have

$$\psi = \sum_{i=1}^{\infty} \psi_i(x) y^{2i-1+\sigma} \tag{4.1}$$

and

$$\rho = \sum_{i=1}^{\infty} \rho_i(x) y^{2i-2}, \tag{4.2}$$

where

$$\psi_i(x) = \sum_{j=1}^{\infty} \psi_{ij}(x-x_1)^{j-1} \tag{4.3}$$

and

$$\rho_i(x) = \sum_{j=1}^{\infty} \rho_{ij}(x-x_1)^{j-1} \tag{4.4}$$

when $|x - x_1|$ and y are sufficiently small. The coefficients $\psi_i(x)$ and $\rho_i(x)$ can be calculated by use of Padé fractions for a given value of x, after which Padé fractions can be formed from (4.1) and (4.2). Padé fractions formed from (4.3) and (4.4) are exact at $x = x_1$, but their accuracy decreases in general as $|x - x_1|$ increases. In a given case, however, it may be possible to find an interval about x_1 throughout which acceptable accuracy is obtained. It would be more economical to use several such intervals with overlapping regions of validity than to solve the inverse problem for each value of x.

In nozzle calculations, expansions in powers of ψ may be more convenient than (4.1) and (4.2). We know by the Cauchy-Kowalewski theorem that for x sufficiently near x_1, (4.1) has a nonzero radius of convergence. Assuming that $\psi_1(x)$ is not zero for a given x, we can invert (4.1) to obtain an expansion of the form

$$y = \sum_{i=1}^{\infty} y_i(x)\, \psi^{2i-1} \tag{4.5}$$

in the two-dimensional case, and

$$y^2 = \sum_{i=1}^{\infty} y_i(x)\, \psi^{i} \tag{4.6}$$

in the axially symmetric case for sufficiently small ψ. By substituting (4.5) or (4.6) in (4.2), we can obtain an expansion for ρ in powers of ψ. Similarly, starting from the Taylor expansion of other flow quantities in the neighborhood of $(x_1, 0)$, we can find their expansions in powers of ψ for constant x.

5. Power Series Along Given Curves

Another procedure for expressing the solution of the inverse problem in terms of power series in one variable is to calculate ψ and ρ along given curves through $(x_1, 0)$. Families of such curves can be chosen which sweep out a neighborhood of $(x_1, 0)$. Let the equation of a given curve be $x - x_1 = g(y)$, where

$$g(y) = \sum_{i=1}^{\infty} g_i y^{i} \tag{5.1}$$

for sufficiently small y. On substituting $x - x_1 = g(y)$ in (2.9) and (2.10), we obtain expansions of the forms

$$\psi = \sum_{i=1}^{\infty} a_i y^{i+\sigma} \tag{5.2}$$

and

$$\rho = \sum_{i=1}^{\infty} b_i y^{i-1}, \tag{5.3}$$

respectively. With the equation of the given curve written in the alternate form $y = g(x - x_1)$, y is replaced by $x - x_1$ in (5.1) through (5.3). In the special case when g(y) is a function of y^2, we have

$$\psi = \sum_{i=1}^{\infty} a_i y^{2i-1+\sigma} \tag{5.4}$$

and

$$\rho = \sum_{i=1}^{\infty} b_i y^{2i-2} \tag{5.5}$$

along the given curve. We see that the coefficients a_i and b_i in (5.2) through (5.5) are known exactly, while the coefficients of (4.1) and (4.2) for $x \neq x_1$ are known only approximately. In addition, $\psi_i(x)$ and $\rho_i(x)$ become less accurate for a given x as i increases, since the number of terms of their Taylor expansions which are available then decreases. However, the accuracy obtained in a particular case is dependent on the choice of x_1, and can be determined only by trial.

As in section 4, if $a_1 \neq 0$, we can invert (5.2) to obtain

$$y = \sum_{i=1}^{\infty} c_i \psi^i \tag{5.6}$$

in the two-dimensional case and

$$y = \sum_{i=1}^{\infty} d_i \psi^{i/2} \tag{5.7}$$

in the axially symmetric case for sufficiently small ψ. Similarly, if $a_1 \neq 0$ in (5.4), we have

$$y = \sum_{i=1}^{\infty} c_i \psi^{2i-1} \tag{5.8}$$

in the two-dimensional case, and

$$y^2 = \sum_{i=1}^{\infty} d_i \psi^i \tag{5.9}$$

in the axially symmetric case. By use of the preceding expansions for y, we can obtain expansions for ρ and other flow quantities along the given curve in powers of ψ or $\psi^{1/2}$.

A special case of an expansion along a curve of the form $x - x_1 = g(y^2)$ is given by the expansions at constant potential of [15]. The equation of the equipotential through the point $(x_1, 0)$ is of the form

$$x - x_1 = \sum_{i=1}^{\infty} p_i y^{2i} \tag{5.10}$$

The calculation of the coefficients p_i, starting from the coefficients of (2.9) and (2.10), is described in detail in [15].

We see that the special choices $g(y) = \alpha y$ and $g(y) = \alpha y^2$ correspond to Moran's procedure in [13]. In [13], a double power series in x and y is written so that terms of the same degree in x and y jointly are grouped together. The given series is then expressed as a power series in one of the variables with coefficients which are polynomials in the ratio of the variables, and this ratio is held constant in a given calculation. The use of the above choices for g(y), after making the substitution $y = z^{1/2}$ in the second, is seen to be equivalent to Moran's procedure.

In the preceding two cases, it is possible to obtain the coefficients a_i and b_i explicitly as polynomials in α. In the case $x - x_1 = \alpha y$ (or $y = \alpha(x - x_1)$), a more symmetrical approach is to transform (2.9) and (2.10) to polar coordinates r and θ with origin at $(x_1, 0)$. On substituting $x - x_1 = r \cos \theta$ and $y^2 = r^2(1 - \cos^2\theta)$ in (2.9) and (2.10), we obtain power series in r for $\psi/y^{1+\sigma}$ and ρ with coefficients which are polynomials in $\cos \theta$.

6. The Equation of the Sonic Line

The equation of the sonic line has an expansion of the form

$$x = \sum_{i=1}^{\infty} s_i y^{2i} \tag{6.1}$$

in the neighborhood of the origin both in the two-dimensional and axially symmetric cases. We can calculate the coefficients s_i successively by substituting (6.1) in the left hand side of the equation

$$\sum_{i=1}^{\infty} \sum_{j=1}^{\infty} \rho_{ij} y^{2i-2} x^{j-1} = 1 \tag{6.2}$$

and equating the coefficients of powers of y^2 past the constant term to zero. This calculation can be carried out on a computer by use of subroutines for power series manipulations. Finally, we can obtain power series for flow quantities along the sonic line by substituting (6.1) or its inversion in their Taylor expansions in the neighborhood of the origin.

7. The Equation of the Limiting Characteristic

As shown in Figure 1, the limiting characteristic is the right-running characteristic which passes through the origin. In both the two-dimensional and axially symmetric cases, the two families of characteristics satisfy the equations

$$dy/dx = \tan(\theta \pm \alpha), \tag{7.1}$$

where $\tan\theta = -\psi_x/\psi_y$ and $\tan\alpha = (M^2 - 1)^{-1/2}$. Making these substitutions, we obtain

$$[-\psi_x(M^2-1)^{1/2} \pm \psi_y]dx/dy - \psi_y(M^2-1)^{1/2} \mp \psi_x = 0. \tag{7.2}$$

The equation of the limiting characteristic has an expansion of the form

$$x = \sum_{i=1}^{\infty} r_i y^{2i} \tag{7.3}$$

for sufficiently small y. Denoting the left hand side of (7.2) by C, we find that

$$C = \sum_{i=1}^{\infty} c_i y^{2i-1+\sigma}. \tag{7.4}$$

On setting $c_1 = 0$ and proceeding as in [15], we obtain

$$r_1 = (\gamma+1)\rho_{12}/4 \tag{7.5}$$

in the two-dimensional case, and

$$r_1 = (\sqrt{5}-1)(\gamma+1)\rho_{12}/8 \tag{7.6}$$

in the axially symmetric case. We note that both values are negative, since ρ_{12} is negative. We can verify that

$$c_i = -[2i - (\gamma+1)\rho_{12}/2d_1]r_i + c_i', \tag{7.7}$$

where c_i' does not involve r_i. Finally, in order to determine r_i for $i \geq 2$, we start from (7.5) in the two-dimensional case and (7.6) in the axially symmetric case, and calculate

$$r_i = c_i'/[2i - (\gamma+1)\rho_{12}/2d_1], \tag{7.8}$$

for $i = 2, 3, \ldots$. For each i, the calculation of c_i' is carried out by setting $r_i = 0$ and calculating c_i.

After the coefficients r_i have been determined, we can find power series for flow quantities along the limiting characteristic by substituting (7.1) or its inversion in their Taylor expansions in the neighborhood of the origin.

8. Numerical Results and Discussion

As in [15], calculations have been carried out on the CDC 6400 in the axially symmetric case with $\gamma = 1.4$, using the velocity on the centerline given by example (c) of [6]. Coefficients of the Taylor expansions (2.9) and (2.10) are computed for $i \leq 25$ in these calculations, and hence, with the maximum value of j equal to 49. The centerline velocity of example (c) is given in terms of the present coordinates and reference quantities by

$$u = A_1 + A_2/[A_3 + (x-A_4)^2]^2, \tag{8.1}$$

where $A_1 = 0.0657267$, $A_2 = 3.1224458$, $A_3 = 1.7125400$, and $A_4 = 0.34000641$. This centerline velocity is shown in Figure 2. In the present units, the value of the stream function which defines the nozzle contour in [6] becomes $\psi = 0.31104$.

Figure 3 shows the calculated nozzle contour, sonic line, and limiting characteristic for the centerline velocity of example (c). Equipotential curves are also shown, as calculated in [15] by means of Padé fractions formed from expansions of $x - x_1$ and y^2 in powers of ψ. Points on the nozzle contour were calculated by use of Padé fractions formed from expansions in powers of y^2 or ψ for $x = x_1$. The sonic line was calculated in two ways, by solving the equation $\rho = 1$ by Newton's method with the left hand side replaced by a Padé fraction, and by forming a Padé fraction from (6.1). In the present example, the former method was found to be more accurate than the latter for $y > 0.6$. Finally, the limiting characteristic was calculated in [15] by means of a Padé fraction formed from the right hand side of (7.3). Comparison is made with the calculations of [6] and [9] by the method of complex characteristics.

Similarly, Figure 4 compares the flow angle calculated in [15] with the calculations of [6] and [9]. The calculations of [15] were carried out by means of Pade´ fractions formed from power series in y^2 for $x = x_1$.

In Tables 1 and 2, Pade´ fractions of the form $f_{n,n}(z)$ are compared with the corresponding partial sums of the series (4.1) for $x = x_1 = -1$ and for two values of y. The tables indicate that the sequence of Pade´ fractions converges for both values of y, while the series converges for the smaller value of y and diverges for the larger. Tables 3 and 4 compare Pade´ fractions and partial sums at the points (-1, 1.1) and (-1, 1.6) for expansions of ψ in powers of y^2 along parabolas of the form $x - x_1 = \alpha y^2$ with $x_1 = -3$. Finally, Tables 5 and 6 give the same comparison for expansions of ψ in powers of y along the rays from (-3,0). The expansions along rays through (-3, 0) contain all powers of y up to and including y^{48}, and hence, 49 coefficients are found in the present calculations. We note that these expansions use all 625 of the coefficients ψ_{ij} obtained in the solution of the inverse problem, while only 325 are used by the expansions along parabolas. The series converges for both values of y in Tables 3 through 6, but more slowly than the sequence of Pade´ fractions.

Comparison of Tables 4 and 6 shows that the expansion along a straight line through (-3, 0) leads to more rapid convergence of the sequence of Pade´ fractions at (-1, 1.6) than the expansion along a parabola, while both tables show more rapid convergence than Table 2. We see that the expansion along $x = -1$ used in Tables 1 and 2 is a special case both of an expansion along a straight line through (-1, 0) and of an expansion along a parabola. It follows that the most suitable center of expansion $(x_1, 0)$ in a particular case is not necessarily the one nearest the point at which the flow is calculated. Further calculations show that the portion of the nozzle contour shown in Figure 3 can be calculated to 4 figures or more for $x < -0.5$ by means of Pade´ fractions formed from power series along rays through (-3, 0). The remaining portion of the nozzle contour in Figure 3 can be calculated by use of Pade´ fractions formed from expansions along rays through the origin and through the point (-0.25, 0). This method is found to be more economical than the procedure of section 4, in which $\psi_i(x)$ and $\rho_i(x)$ are calculated by means of Pade fractions, since the range of x for which the latter are sufficiently accurate becomes small for $i \geq 15$ in the present calculations. The time required to find expansions along a given ray through $(x_1, 0)$ when ψ_{ij} and ρ_{ij} are known is much less than that for solution of the inverse problem.

9. References

1. J. R. Baron, "Analytic Design of a Family of Supersonic Nozzles by the Friedrichs Method," WADC Report 54-279, June 1954, Naval Supersonic Lab., MIT, Cambridge, Mass.

2. D. F. Hopkins and D. E. Hill, "Effect of Small Radius of Curvature on Transonic Flow in Axisymmetric Nozzles," AIAA Journal, Vol. 4 (1966), pp. 1337-1343.

3. J. D. Anderson, Jr. and E. L. Harris, "Modern Advances in the Physics of Gasdynamic Lasers," AIAA Paper 72-143, San Diego, Calif., 1972.

4. T. A. Cool, "A Summary of Recent Research on Continuous Wave Chemical Lasers," Modern Optical Methods in Gas Dynamics Research, edited by D. S. Dosanjh, Plenum Press, New York, 1971, pp. 197-220.

5. J. V. Armitage, "Flow in a deLaval Nozzle by the Garabedian Method," ARL 66-0012, Jan. 1966, Aerospace Research Labs., Dayton, Ohio.

6. G. V. R. Rao and B. Jaffe, "A Numerical Solution of Transonic Flow in a Convergent-Divergent Nozzle," Final Report, Contract NAS7-635, March 1969, NASA.

7. P. R. Garabedian, "Numerical Construction of Detached Shock Waves," Journal of Mathematics and Physics, Vol. 36 (1957), pp. 192-205.

8. P. R. Garabedian, Partial Differential Equations, Wiley, New York, 1964, Chapter 16.

9. J. M. Solomon, private communication, Jan. 1971, Naval Surface Weapons Center, White Oak, Silver Spring, Md.

10. R. Courant and D. Hilbert, Methods of Mathematical Physics, Vol. II, Interscience Publishers, New York, 1962.

11. M. D. Van Dyke, "A Model of Supersonic Flow Past Blunt Axisymmetric Bodies with Application to Chester's Solution," Journal of Fluid Mechanics, Vol. 3 (1958), pp. 515-522.

12. A. H. Van Tuyl, "The Use of Rational Approximations in the Calculation of Flows With Detached Shocks," Journal of the Aero/Space Sciences, Vol. 27 (1960), pp. 559-560.

13. J. P. Moran, "Initial Stages of Axisymmetric Shock-on-Shock Interaction for Blunt Bodies," Physics of Fluids, Vol. 13 (1970), pp. 237-248.

14. A. H. Van Tuyl, "Use of Padé Fraction in the Calculation of Blunt Body Flows," AIAA Journal, Vol. 9 (1971), pp. 1431-1433.

15. A. H. Van Tuyl, "Calculation of Nozzle Flows Using Padé Fractions," AIAA Journal, Vol. 11 (1973), pp. 537-541.

16. H. S. Wall, Analytic Theory of Continued Fractions, D. Van Nostrand, New York, 1948.

17. P. Henrici, "The Quotient-Difference Algorithm," Further Contributions to the Solution of Simultaneous Linear Equations and the Determination of Eigenvalues, National Bureau of Standards Applied Mathematics Series, No. 49, 1958, pp. 23-46.

18. P. Wynn, "On a Device for Computing the $e_m(S_n)$ Transform," Math. Tables and Other Aids to Computation, Vol. 10 (1956), pp. 91-96.

Table 1. Padé fractions for stream function at (-1, 1.1) from power series on the line x = -1.

No. of terms	Series	Padé fractions
11	0.22136847	0.22138164
13	0.22139025	0.22138385
15	0.22137986	0.22138366
17	0.22138595	0.22138367
19	0.22138235	0.22138366
21	0.22138439	0.22138367
23	0.22138328	0.22138367
25	0.22138386	0.22138367

Table 2. Padé fractions for stream function at (-1, 1.6) from power series on the line x = -1.

No. of terms	Series	Padé fractions
11	0.86601815	0.32739899
13	0.72548523	0.32960800
15	-0.44164134	0.32908139
17	1.9872070	0.32913508
19	-3.3585190	0.32907155
21	8.5084977	0.32911728
23	-17.484076	0.32912434
25	38.076299	0.32911631

Table 3. Pade´ fractions for stream function at (-1, 1.1) from power series along parabola through (-3, 0).

No. of terms	Series	Pade´ fractions
11	0.22082834	0.22139959
13	0.22121951	0.22139261
15	0.22135508	0.22140720
17	0.22138460	0.22138378
19	0.22138868	0.22138298
21	0.22138819	0.22138362
23	0.22138670	0.22138367
25	0.22138537	0.22138366

Table 4. Pade´ fractions for stream function at (-1. 1.6) from power series along parabola through (-3, 0).

No. of terms	Series	Pade´ fractions
11	0.33051685	0.32899479
13	0.32947504	0.32910273
15	0.32910476	0.32913379
17	0.32904116	0.32911671
19	0.32906254	0.32911675
21	0.32909083	0.32912158
23	0.32910815	0.32911528
25	0.32911519	0.32911678

Table 5. Pade´ fractions for stream function at (-1, 1.1) from power series along straight line through (-3, 0).

No. of terms	Series	Pade´ fractions
35	0.22138349	0.22138363
37	0.22138358	0.22138366
39	0.22138358	0.22138367
41	0.22138363	0.22138367
43	0.22138366	0.22138367
45	0.22138366	0.22138367
47	0.22138366	0.22138367
49	0.22138367	0.22138367

Table 6. Pade´ fractions for stream function at (-1, 1.6) from power series along straight line through (-3, 0).

No. of terms	Series	Pade´ fractions
35	0.32910043	0.32913355
37	0.32910691	0.32911442
39	0.32910959	0.32911469
41	0.32911685	0.32911464
43	0.32911775	0.32911467
45	0.32911594	0.32911474
47	0.32911575	0.32911480
49	0.32911506	0.32911470

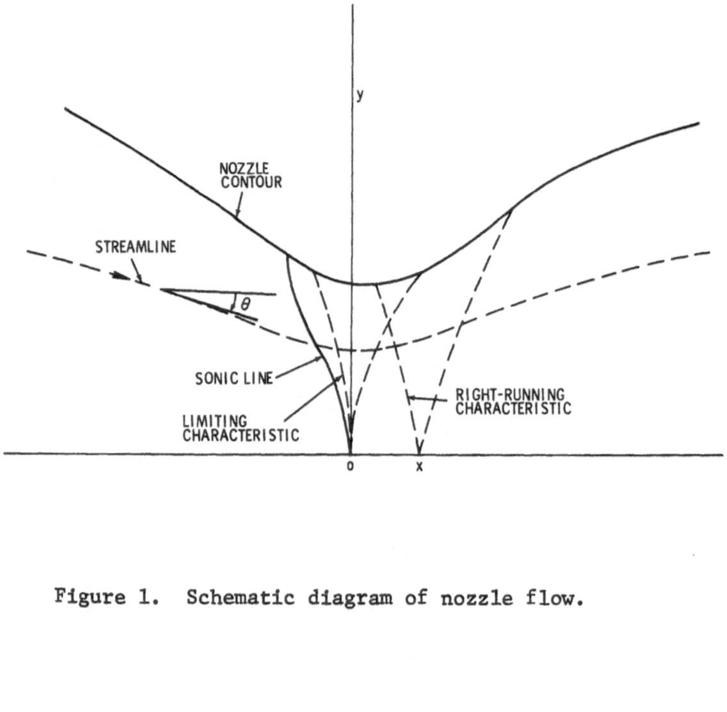

Figure 1. Schematic diagram of nozzle flow.

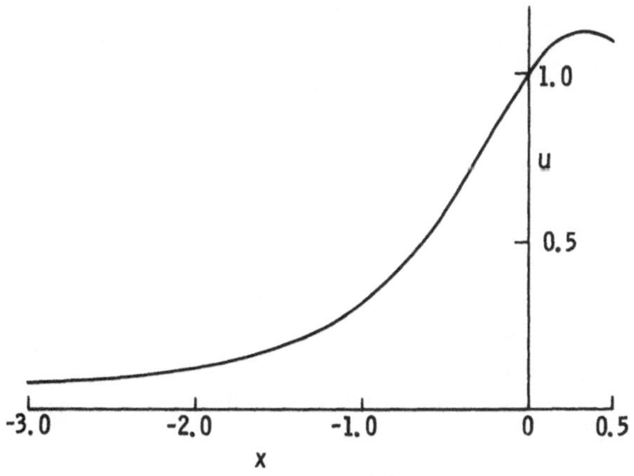

Figure 2. Centerline velocity distribution of example (c) of [6].

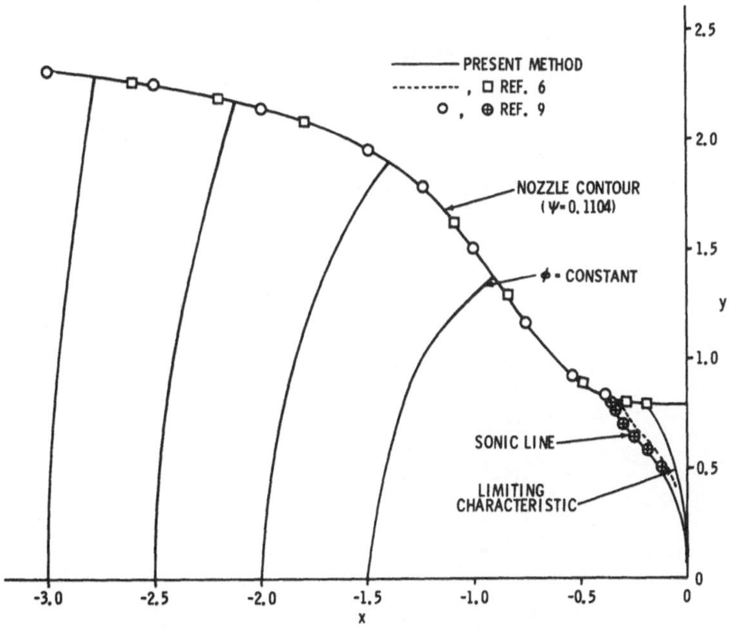

Figure 3. Subsonic and transonic portions of nozzle in example (c).

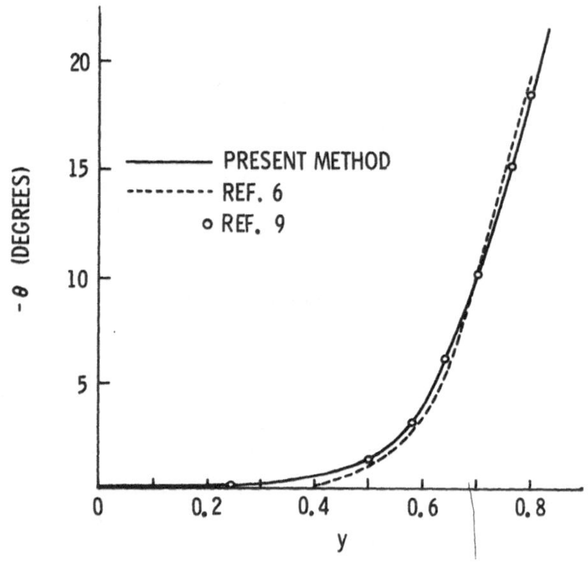

Figure 4. Flow angle θ along sonic line in example (c).

Part III

General Bibliography

A BIBLIOGRAPHY ON PADE APPROXIMATION AND SOME RELATED MATTERS

Claude BREZINSKI

University of Lille

The aim of this paper is to give a bibliography on Padé approximants, some related matters and applications.

For several years, Padé approximants had become more and more important in mathematics, numerical analysis and various fields in physics and engineering. They are closely related to many subjects in mathematics as analytic function theory, difference equations, the theory of moments, approximation, analytic continuation, continued fractions, etc. Then a whole bibliography should be a huge one to include the corresponding references of these disciplines.

I have divided the references given in this paper into three sections. The first one deals with Padé approximation and I hope it is quite complete. The second one is devoted to continued fractions and includes only some historical references and most of the recent papers on this subject. The thrid section contains some applications of Padé approximants with a special emphasis on mechanics ; I have also included references on numerical analysis and methods to accelerate the convergence of sequences. Miscellaneous references end the paper.

It is obvious that this bibliography is far to be complete because of the limited number of pages of this volume. It is, in fact, less than half of a bigger bibliography on this subject and on all the related matters that I hope to publish in the future. I apologize in advance for any errors and omissions and I thank everybody who would send me any new reference on this subject.

I - PADE approximants

1 - R.J. ARMS, A. EDREI - The Padé tables and continued fractions generated by totally positive sequences - in "Mathematical essays dedicated to A.J. Macintyre" (1970) 1-21.

2 - G.A. BAKER Jr.- The Padé approximant method and some related generalizations - in "The Padé approximant in theoretical physics", G.A. Baker Jr. and J.L. Gammel eds., Academic Press, New York, 1970.

3 - G.A. BAKER Jr.- The theory and application of the Padé approximant method - J. Adv. Theor. Phys., 1 (1965) 1-56.

4 - G.A. BAKER Jr.- Recursive calculation of Padé approximants - in "Padé approximants and their applications", P.R. Graves - Morris ed., Academic Press, New-York, 1973.

5 - G.A. BAKER Jr.- Best error bounds for Padé approximants to convergent series of Stieltjes - J. Math. Phys., 10 (1969) 814-820.

6 - G.A. BAKER Jr.- Certain invariance and convergence properties of Padé approximants - Rocky Mountains J. Math., 4 (1974) 141-150.

7 - G.A. BAKER Jr.- The existence and convergence of subsequences of Padé approximants - J. Math. Anal. Appl., 43 (1973) 498-528.

8 - G.A. BAKER Jr.- Essential of Padé approximants - Academic Press, New-York, 1975.

9 - G.A. BAKER Jr., J.L. GAMMEL eds. - The Padé approximant in theoretical physics- Academic Press, New York, 1972.

10 - G.A. BAKER Jr., J.L. GAMMER, J.G. WILLS - An investigation of the applicability of the Padé approximant method - J. Math. Anal. Appl., 2 (1961) 405-418.

11 - M. BARNSLEY - The bounding properties of the multipoint Padé approximant to a series of Stieltjes - Rocky Mountains J. Math., 4 (1974) 331-334.

12 - M. BARNSLEY - The bounding properties of the multipoint Padé approximant to a series of Stieltjes on the real line - J. Math. Phys., 14 (1973) 299-313.

13 - M. BARNSLEY, P.D. ROBINSON - Dual variational principles and Padé-type approximants - J. Inst. Math. Appl., 14 (1974) 229-250.

14 - M. BARNSLEY, P.D. ROBINSON - Padé approximant bounds and approximate solutions for Kirkwood - Riseman integral equations - J. Inst. Math. Appl., 14 (1974) 251-265.

15 - J.L. BASDEVANT - Padé approximants - in "Methods in subnuclear physics", Vol. IV, Gordon and Breach, London, 1970.

16 - J.L. BASDEVANT - The Padé approximation and its physical applications - Fort. der Physik, 20 (1972) 283-331.

17 - A.F. BEARDON - On the convergence of Padé approximants - J. Math. Anal. Appl., 21 (1968) 344-346.

18 - D. BESSIS - Topics in the theory of Padé approximants - in "Padé approximants", P.R. Graves-Morris ed., The institute of physics, London, 1973.

19 - D. BESSIS, J.D. TALMAN - Variational approach to the theory of operator Padé approximants - Rocky Mountains J. Math., 4 (1974) 151-158.

20 - C. BREZINSKI - Convergence of Padé approximants for some special sequences - Thrid Colloquium on advanced computing methods in theoretical physics, Marseille, 1973.

21 - C. BREZINSKI - Rhombus algorithms connected to the Padé table and continued fractions - to appear.

22 - C. BREZINSKI - Computation of Padé approximants and continued fractions - to appear.

23 - C. BREZINSKI - Séries de Stieltjes et approximants de Padé - Colloque Euromech 58, Toulon, 12-14 mai 1975.

24 - C. BREZINSKI - Some results in the theory of the vector ε-algorithm - Linear Algebra, 8 (1974) 77-86.

25 - C. BREZINSKI - Some results and applications about the vector ε-algorithm - Rocky Mountains J. Math., 4 (1974) 335-338.

26 - C. BREZINSKI - Généralisations de la transformation de Shanks, de la table de Padé et de l'ε-algorithme - Calcolo (to appear).

27 - J.S.R. CHISHOLM - Mathematical theory of Padé approximants - in "Padé approximants", P.R. Graves-Morris ed., The institute of Physics, London, 1973.

28 - J.S.R. CHISHOLM - Rational approximants defined from double power series - Math. Comp., 27 (1973) 841-848.

29 - J.S.R. CHISHOLM - Approximation by sequences of Padé approximants in regions of meromorphy - J. Math. Phys., 7 (1966) 39-44.

30 - J.S.R. CHISHOLM - Convergence properties of Padé approximants - in "Padé approximants and their applications", P.R. Graves-Morris ed., Academic Press, New-York, 1973.

31 - C.K. CHUI, O. SHISHA, P.W. SMITH - Padé approximants as limits of best rational approximants - J. Approx. Theory, 12 (1974) 201-204.

32 - A.K. COMMON - Padé approximants and bounds to series of Stieltjes - J. Math. Phys., 9 (1968) 32-38.

33 - A.K. COMMON, P.R. GRAVES-MORRIS - Some properties of Chisholm approximants - J. Inst. Math. Applics., 13 (1974) 229-232.

34 - J.D.P. DONNELLY - The Padé table - in "Methods of numerical approximation", D.C. Handscomb ed., Pergamon Press, New-York, 1966.

35 - A. EDREI - The Padé table of meromorphic functions of small order with negative zeros and positive poles - Rocky Mountains J. Math., 4 (1974) 175-180.

36 - A. EDREI - Convergence de la méthode de Padé appliquées aux fonctions méromorphes - Colloque Euromech 58, Toulon, 12-14 mai 1975.

37 - D. ELLIOT - Truncation errors in Padé approximations to certain functions : an alternative approach - Math. Comp., 21 (1967) 398-406.

38 - S.T. EPSTEIN, M. BARNSLEY - A variational approach to the theory of multipoint Padé approximants - J. Math. Phys., 14 (1973) 314-325.

39 - W. FAIR - Padé approximation to the solution of the Riccati equation - Math. Comp., 18 (1964) 627-634.

40 - W. FAIR , Y.L. LUKE - Padé approximations to the operator exponential - Numer. Math., 14 (1970) 379-382.

41 - J. FLEISCHER - Nonlinear Padé approximants for Legendre series - J. Math. Phys., 14 (1973) 246-248.

42 - J. FLEISCHER - Nonlinear Padé approximants for Legendre series - in "Padé approximants and their applications", P.R. Graves-Morris ed., Academic Press, New-York, 1973.

43 - J. FLEISCHER - Generalizations of Padé approximants - in "Padé approximants", P.R. Graves-Morris ed., The institute of physics, London, 1973.

44 - N.R. FRANZEN - Some convergence results for Padé approximants - J. Approx. Theory, 6 (1972) 254-263.

45 - N.R. FRANZEN - Convergence of Padé approximants for a certain class of meromorphic functions - J. Approx. Theory, 6 (1972) 264-271.

46 - J.L. GAMMEL - Review of two recent generalizations of the Padé approximant - in "Padé approximants and their applications", P.R. Graves-Morris ed., Academic Press, New-York, 1973.

47 - J.L. GAMMEL - Effect of random errors (noise) in the terms of a power series on the convergence of the Padé approximants - in "Padé approximants", P.R. Graves-Morris ed., The institute of physics, London, 1973.

48 - J.L. GAMMEL, J. NUTTALL - Convergence of Padé approximants to quasi-analytic functions beyond natural boundaries - J. Math. Anal. Appl.

49 - J. GILEWICZ - Numerical detection of the best Padé approximant and determination of the Fourier coefficients of the insufficiently sampled functions - in "Padé approximants and their applications", P.R. Graves-Morris ed., Academic Press, New-York, 1973.

50 - J. GILEWICZ - Totally monotonic and totally positive sequences for the Padé approximants method - Colloque Euromech 58, Toulon, 12-14 mai 1975.

51 - J.E. GOLDEN, J.H. McGUIRE, J. NUTTALL - Calculating Bessel functions with Padé approximants - J. Math. Anal. Appl., 43 (1973) 754-767.

52 - W.B. GRAGG - The Padé table and its relation to certain algorithms of numerical analysis - SIAM Rev., 14 (1972) 1-62.

53 - W.B. GRAGG - On Hadamar's theory of polar singularities - in "Padé approximants and their applications", P.R. Graves-Morris ed., Academic Press, New-York, 1973.

54 - W.B. GRAGG, G.D. JOHNSON - The Laurent - Padé table - Proceedings IFIP Congress, North-Holland, 1974.

55 - P.R, GRAVES-MORRIS ed. - Padé approximants and their applications- Academic Press, New-York, 1973.

56 - P.R. GRAVES-MORRIS ed. - Padé approximants - The institute of physics, London, 1973.

57 - T.N.E. GREVILLE - On some conjectures of P. Wynn concerning the ε-algorithm - MRC Technical summary report 877, Madison, 1968.

58 - A.S. HOUSEHOLDER - The Padé table, the Frobenius identities and the q-d algorithm - Linear Algebra, 4 (1971) 161-174.

59 - A.S. HOUSEHOLDER, G.W. STEWART - Bigradients, Hankel determinants and the Padé table - in "Constructive aspects of the fundamental theorem of algebra", B. Dejon and P. Henrici eds., Academic Press, New-York, 1969.

60 - R.C. JOHNSON - Alternative approach to Padé approximants - in "Padé approximants and their application", P.R. Graves-Morris ed., Academic Press, New-York, 1973.

61 - W.B. JONES - Truncation error bound for continued fractions and Padé approximants - in "Padé approximants and their applications", P.R. Graves-Morris ed., Academic Press, New-York, 1973.

62 - W.B. JONES - Analysis of truncation error of approximations based on the Padé table and continued fractions - Rocky Mountains J. Math., 4 (1974) 241-250.

63 - W.B. JONES, W.J. THRON - On convergence of Padé approximants - SIAM J. Math. Anal., 6 (1975) 9-16.

64 - I.M. LONGMAN - Computation of the Padé table - Intern. J. Comp. Math., 3B (1971) 53-64.

65 - Y.L. LUKE - Evaluation of the gamma function by means of Padé approximations - SIAM J. Math. Anal., 1 (1970) 266-281.

66 - Y.L. LUKE - The Padé table and the τ-method - J. Math. Phys., 37 (1958) 110-127.

67 - A. MAGNUS - Certain continued fractions associated with the Padé table - Math. Z., 78 (1960) 361-374.

68 - A. MAGNUS - P-fractions and the Padé table - Rocky Mountains J. Math., 4 (1974) 257-260.

69 - D. MASSON - Hilbert space and Padé approximant - in "The Padé approximant in theoretical physics", G.A. Baker Jr. and J.L. Gammel eds., Academic Press, New-York, 1970.

70 - D. MASSON - Padé approximants and Hilbert spaces - in "Padé approximants and their applications", P.R. Graves-Morris ed., Academic Press, 1973.

71 - J.H. McCABE - A formal extension of the Padé table to include two point Padé quotients - J. Inst. Math. Applics., 15 (1975) 363-372.

72 - J.B. McLEOD - A note on the ε-algorithm - Computing, 7 (1971) 17-24.

73 - G. MERZ - Padésche Näherungsbrücke und Iterationsverfahren höheren Ordnung - Doctoral thesis, Technische Hochschule Clausthal, Germany, 1967.

74 - J. NUTTALL - The connection of Padé approximants with stationary variational principles and the convergence of certain Padé approximants - in "The Padé approximant in theoretical physics", G.A. Baker Jr. and J.L. Gammel eds., Academic Press, New-York, 1970.

75 - J. NUTTALL - The convergence of Padé approximants of meromorphic functions - J. Math. Anal. Appl., 31 (1970) 147-153.

76 - J. NUTTALL - Variational principles and Padé approximants - in "Padé approximants and their applications", P.R. Graves-Morris ed., Academic Press, New-York, 1973.

77 - J. NUTTALL - The convergence of certain Padé approximants - Rocky Mountains J. Math., 4 (1974) 269-272.

78 - H. PADE - Sur la représentation approchée d'une fonction par des fractions rationnelles - Ann. Ec. Norm. Sup., 9 (1892) 1-93.

79 - C. POMMERENKE - Padé approximants and convergence in capacity - J. Math. Anal. Appl., 41 (1973) 775-780.

80 - L.D. PYLE - A generalized inverse ε-algorithm for constructing intersection projection matrices, with applications - Numer. Math., 10 (1967) 86-102.

81 - J. RISSANEN - Recursive evaluation of Padé approximants for matrix sequences-

IBM J. Res. develop., (july 1972) 401-406.

82 - E.B. SAFF - An extension of Montessus de Ballore's theorem on the convergence of interpolating rational functions - J. Approx. Theory, 6 (1972) 63-67.

83 - W.F. TRENCH - An algorithm for the inversion of finite Hankel matrices - SIAM J. Appl. Math., 13 (1965) 1102-1107.

84 - R.P. WAN DE RIET - On certain Padé polynomials in connection with the block function - report TN 32, Mathematical center, Amsterdam, 1963.

85 - H. VAN ROSSUM - A theory of orthogonal polynomials based on the Padé table - Van Goraim, Assen, 1953.

86 - H.S. WALL - On the relationship among the diagonal files of a Padé table - Bull. Amer. Math. Soc., 38 (1932) 752-760.

87 - H.S. WALL - On the Padé approximants associated with the continued fraction and series of Stieltjes - Trans. Amer. Math. Soc., 31 (1929) 91-116.

88 - H.S. WALL - On the Padé approximants associated with a positive definite power series - Trans. Amer. Math. Soc., 33 (1931) 511-532.

89 - H.S. WALL - General theorems on the convergence of sequences of Padé approximants - Trans. Amer. Math. Soc., 34 (1932) 409-416.

90 - H. WALLIN - On the convergence theory of Padé approximants - Proceedings of the Oberwolfach conference on approximation theory, 1971.

91 - H. WALLIN - The convergence of Padé approximants and the size of the power series coefficients - Applicable Anal., 4 (1974) 235-252.

92 - J.L. WALSH - Padé approximants as limit of rational functions of best approximation - J. Math. Mech., 13 (1964) 305-312.

93 - J.L. WALSH - Padé approximants as limits of rational functions of best approximation, real domain - J. Approx. Theory, 11 (1974) 225-230.

94 - P. WYNN - Upon the inverse of formal power series over certain algebras - Centre de recherches mathématiques, Université de Montréal, 1970.

95 - P. WYNN - Upon the generalized inverse of a formal power series with vector valued coefficients - Compositio Math., 23 (1971) 453-460.

96 - P. WYNN - Sur l'équation aux dérivées partielles de la surface de Padé - C.R. Acad. Sc. Paris, 278 A (1974) 847-850.

97 - P. WYNN - Uber finen interpolations - algorithmus und gewise andere formeln, die in der theorie der interpolation durch rationale funktionen bestehen - Numer. Math., 2 (1961) 151-182.

98 - P. WYNN - Difference - differential recursions for Padé quotients - Proc. London Math. Soc., 23 (1971) 283-300.

99 - P. WYNN - A general system of orthogonal polynomials - Quart. J. Math., 18 ser. 2 (1967) 69-81.

100 - P. WYNN - Some recent developments in the theories of continued fractions

and the Padé table - Rocky Mountains J. Math., 4 (1974) 297-324.

101 - P. WYNN - Upon the diagonal sequences of the Padé table - MRC technical summary report 660, Madison, 1966.

102 - P. WYNN - Extremal properties of Padé quotients - Acta Math. Acad. Sci. Hungaricae, 25 (1974) 291-298.

103 - P. WYNN - Upon a convergence result in the theory of the Padé table - Trans. Amer. Math. Soc., 165 (1972) 239-249.

104 - P. WYNN - Zur theorie der mit gewissen speziellen funktionen verknüpften Padèschen tafeln - Math. Z., 109 (1969) 66-70.

105 - P. WYNN - Upon the Padé table derived from a Stieltjes series - SIAM J. Numer. Anal., 5 (1968) 805-834.

106 - P. WYNN - Upon systems of recursions which obtain among the quotients of the Padé table - Numer. Math., 8 (1966) 264-269.

107 - P. WYNN - L'ε-algoritmo e la tavola di Padé - Rend. di Mat. Roma, 20 (1961) 403-408.

108 - P. WYNN - Upon a conjecture concerning a method for solving linear equations, and certain other matters - MRC technical summary report 626, Madison, 1966.

109 - P. WYNN - General purpose vector epsilon algorithm procedures - Numer. Math., 6 (1964) 22-36.

110 - P. WYNN - Acceleration techniques for iterated vector and matrix problems - Math. Comp., 16 (1962) 301-322.

II - Continued fractions

111 - F.L. BAUER - Use of continued fractions and algorithms related to them - CIME summer school lectures, Perigia, 1964.

112 - F.L. BAUER, E. FRANK - Note on formal properties of certain continued fractions - Proc. Amer. Math. Soc., 9 (1958) 340-347.

113 - D. BERNOULLI - Adversaria analytica miscellanea de fractionibus continuis - Novi comm. Acad. Sc. Imp. St Pétersbourg, 20 (1775)

114 - D. BERNOULLI - Disquisitiones ulteriores de indole fractionum continuarum - Novi comm. Acad. Sc. Imp. St Pétersbourg, 20 (1775)

115 - S. BERNSTEIN, O. SZASZ - Über Irrationalität unendlicher Kettenbrüche mit einer Anwendung auf die Reihe $\Sigma\, q^{\nu^2} x^{\nu}$ - Math. Ann., 76 (1915) 295-300.

116 - G. BLANCH - Numerical evaluation of continued fractions - SIAM Rev., 7 (1964) 383-421.

117 - E. CESARO - Sur quelques fractions continues - Nouvelles Ann. de Math., (3) 6 (1887)

118 - P. CHEBYCHEV - Sur les fractions continues - Jour. de Math., ser. II,3

(1858) 289-323.

119 - V.F. COWLING, W. LEIGHTON, W.J. THRON - Twin convergence regions for the general continued fraction - Bull. Amer. Math. Soc., 49 (1943) 913-916.

120 - I.V. CYGANKOV - Solution of Riccati equations by continued fractions - Perm. Gos. Univ. Ucen. Zap. Mat., 17 (1960) 99-107.

121 - I.V. CYGANKOV - Solution of a special Riccati equation by continued fractions - Perm. Gos. Univ. Ucen. Zap. Mat., 17 (1960) 109-113.

122 - J.D.P. DONNELLY - Continued fractions - in "Methods of numerical approximation", D.C. Handscomb ed., Pergamon Press, Oxford, 1965.

123 - H.G. ELLIS - Continued fractions solutions of the general Riccati differential equation - Rocky Mountains J. Math., 4 (1974) 353-356.

124 - L. EULER - De fractionibus continuis observationes - Comm. Acad. Sc. Imp. St Pétersbourg, 11 (1739).

125 - L. EULER - De formatione fractionum continuarum - Acta Acad. Sc. Imp. St Pétersbourg, 1779.

126 - W. FAIR - Noncommutative continued fractions - SIAM J. Math. Anal., 2 (1971) 226-232.

127 - W. FAIR - A convergence theorem for noncommutative continued fractions - J. Approx. Theory, 5 (1972) 74-76.

128 - W. FAIR - Continued fraction solution to the Riccati equation in a Banach algebra - J. Math. Anal. Appl., 39 (1972) 318-323.

129 - W. FAIR - Continued fraction solution to Fredholm integral equations - Rocky Mountains J. Math., 4 (1974) 357-360.

130 - D.A. FIELD - A priori truncation error estimates for continûed fractions $K(1/b_n)$ - Rocky Mountains J. Math., 4 (1974) 361-362.

131 - D.A. FIELD, W.B. JONES - A priori estimates for truncation error of continued fractions $K(1/b_n)$ - Numer. Math., 19 (1972) 283-302.

132 - E. FRANK - Corresponding type continued fractions - Amer. J. Math., 68 (1946) 89-108.

133 - E. GALOIS - Démonstration d'un théorème sur les fractions continues périodiques - Ann. Math. Pures et Appl., 19 (1828-1829)

134 - H.L. GARABEDIAN, H.S. WALL - Topics in continued fractions and summability - Northwestern Univ. Studies in Math. and phys. sciences, Vol. 1, Evanston and Chicago, (1941) 89-132.

135 - I. GARGANTINI, P. HENRICI - A continued fraction algorithm for the computation of higher transcendental function in the complex plane - Math. Comp., 21 (1967) 18-29.

136 - T.F. GLASS, W. LEIGHTON - On the convergence of a continued fraction - Bull. Amer. Math. Soc., 49 (1943) 133-135.

137 - W.B. GRAGG - Truncation error bounds for π-fractions - Bull. Amer. Math.

Soc., 76 (1970) 1091-1094.

138 - W.B. GRAGG - Matrix interpretations and applications of the continued fractions algorithm - Rocky Mountains J. Math., 4 (1974) 213-226.

139 - W.B. GRAGG - Truncation error bounds for g-fractions - Numer. Math., 11 (1968) 370-379.

140 - H. HAMBURGER - Ueber di Konvergenz eines mit einer Potenzreihe assoziierten Kettenbruchs - Math. Ann., 81 (1920) 31-45.

141 - H. HAMBURGER - Beiträge zur Konvergenztheorie der Stieltjes ' schen Kettenbrüche - Math. Zeit., 4 (1919) 186-222.

142 - T.L. HAYDEN - Continued fraction approximation to functions - Numer. Math., 7 (1965) 292-309.

143 - T.L. HAYDEN - Continued fractions in Banach spaces - Rocky Mountains J. Math., 4 (1974) 367-370.

144 - T.L. HAYDEN - A convergence problem for continued fractions - Proc. Amer. Math. Soc., 14 (1963) 546-552.

145 - E. HELLINGER - Zur Stieltjesschen Kettenbruchtheorie - Math. Ann., 86 (1922) 18-29.

146 - P. HENRICI - Error bounds for computations with continued fractions - in "Error in digital computation", Wiley, New-York, 1965.

147 - P. HENRICI, P. PFLUGER - Truncation error estimates for Stieltjes fractions - Numer. Math., 9 (1966) 120-138.

148 - K.L. HILLAM - Some convergence criteria for continued fractions - Doctoral thesis, University of Colorado, Boulder, 1962.

149 - K.L. HILLAM, W.J. THRON - A general convergence criterion for continued fractions $K(a_n/b_n)$ - Proc. Amer. Math. Soc., 16 (1965) 1256-1262.

150 - A. HURWITZ - Über eine besondere Art der Kettenbruchentwicklung reeller Grössen - Acta Math., 12 (1889) 367-405.

151 - C.G.J. JACOBI - De fractione continue, in quam integrale $\int_x^\infty e^{-x^2} dx$ evoldere licet - J. für die reine u. angew. math., 12 (1834), 346-347.

152 - T.H. JEFFERSON - Truncation error estimates for T-fractions - SIAM J. Numer. Anal., 6 (1969) 359-364.

153 - W.B. JONES - Contributions to the theory of Thron continued fractions - Ph. D. thesis, Vanderbilt University, Nashville, Tennessee, 1963.

154 - W.B. JONES, R.I. SNELL - Truncation error bounds for continued fractions - SIAM J. Numer. Anal., 6 (1969) 210-221.

155 - W.B. JONES, R.I. SNELL - Sequences of convergence regions for continued fractions $K(a_n/1)$ - Trans. Amer. Math. Soc., 170 (1972) 483-497.

156 - W.B. JONES, W.J. THRON - Convergence of continued fractions - Canad. J. Math., 20 (1968) 1037-1055.

157 - W.B. JONES, W.J. THRON - Twin convergence regions for continued fractions $K(a_n/1)$ - Trans. Amer. Math. Soc., 150 (1970) 93-119

158 - W.B. JONES, W.J. THRON - A posteriori bounds for the truncation error of continued fractions - SIAM J. Numer. Anal., 8 (1971) 693-705.

159 - W.B. JONES, W.J. THRON - Numerical stability in evaluating continued fractions - Math. Comp., 28 (1974) 795-810.

160 - W.B. JONES, W.J. THRON - Further properties of T-fractions - Math. Ann., 166 (1966) 106-118.

161 - J.Q. JORDAN, W. LEIGHTON - On the permutation of the convergents of a continued fraction and related convergence criteria - Ann. Math., (2) 39 (1938) 872-882.

162 - A. Ya. KHINTCHINE - Continued fractions - P. Noordhoff, Groningen, 1963.

163 - A.N. KHOVANSKII - The application of continued fractions and their generalizations to problems in approximation theory - P. Noordhoff, Groningen, 1963.

164 - E. LAGUERRE - Sur la réduction en fractions continues d'une classe assez étendue de fonctions - C.R. Acad. Sc. Paris, 87A (1878).

165 - E. LAGUERRE - Sur la réduction en fractions continues de $e^{F(x)}$, $F(x)$ désignant un polynôme entier - J. Math. pures et appl., (3) 6 (1880).

166 - R.E. LANE - The value region problem for continued fractions - Duke Math. J., 12 (1945) 207-216.

167 - R.E. LANE - The convergence and values of periodic continued fractions - Bull. Amer. Math. Soc., 51 (1945) 246-250.

168 - L.J. LANGE - On a family of twin convergence regions for continued fractions - Ill. J. of Math., 10 (1966) 97-108.

169 - L.J. LANGE, W.J. THRON - A two-parameter family of best twin convergence regions for continued fractions - Math. Zeit., 73 (1969) 277-282.

170 - W. LEIGHTON - A test-ratio test for continued fractions - Bull. Amer. Math. Soc., 45 (1939) 97-100.

171 - W. LEIGHTON - Convergence theorems for continued fractions - Duke Math. J., 5 (1939) 298-308.

172 - W. LEIGHTON - Sufficient conditions for the convergence of a continued fraction - Duke Math. J., 4 (1938) 775-778.

173 - W. LEIGHTON, W.T. SCOTT - A general continued fraction expansion - Bull. Amer. Math. Soc., 45 (1939) 596-605.

174 - W. LEIGHTON, W.J. THRON - On value regions for continued fractions - Bull. Amer. Math. Soc., 48 (1942) 917-920.

175 - W. LEIGHTON, W.J. THRON - Continued fractions with complex elements - Duke Math. J., 9 (1942) 763-772.

176 - W. LEIGHTON, W.J. THRON - On the convergence of continued fractions to

meromorphic functions - Ann. Math., (2) 44 (1943) 80-89.

177 - J.S. MAC NERMEY - Investigations concerning positive definite continued fractions - Duke Math. J., 26 (1959) 663-678.

178 - A. MAGNUS - Expansion of power series into P-fractions - Math. Z., 80 (1962) 209-216.

179 - A. MAGNUS - On P-expansions of power series - Det. Kgl. Norske Videnskabers Selskabs Skrifter (1964), n°3.

180 - A. MAGNUS - The connection between P-fractions and associated fractions - Proc. Amer. Math. Soc., 25 (1970) 676-679.

181 - A. MARKOV - Deux démonstrations de la convergence de certaines fractions continues - Acta Math., 19 (1895) 93 - 104.

182 - A. MARKOV - Note sur les fractions continues - Bull. classe Physico-Math. Acad. Imp. Sc. St Pétersbourg, 5 (2) (1895) 9-13.

183 - D.F. MAYERS - Economization of continued fractions - in "Methods of numerical approximation", D.C. Handscomb ed., Pergamon Press, Oxford, 1965.

184 - J.H. Mc CABE - A continued fraction expansion, with a truncation error estimate, for Dawson's integral - Math. Comp., 28 (1974) 811-816.

185 - E.P. MERKES - On truncation errors for continued fraction computations - SIAM J. Numer. Anal., 3 (1966) 486-496.

186 - E.P. MERKES, W.T. SCOTT - Continued fraction solutions of the Riccati equation - J. Math. Anal. Appl., 4 (1962) 309-327.

187 - R. DE MONTESSUS DE BALLORE - Sur les fractions continues algébriques - Bull. Soc. Math. de France, 30 (1902) 28-36.

188 - R. DE MONTESSUS DE BALLORE - Sur les fractions continues algébriques - C.R. Acad. Sc. Paris, 134 (1902) 1489-1491.

189 - R. DE MONTESSUS DE BALLORE - Sur les fractions continues algébriques de Laguerre - C.R. Acad. Sc. Paris, 140 (1905) 1438-1440.

190 - R. DE MONTESSUS DE BALLORE - Les fractions continues algébriques - Acta Math., 32 (1909) 257-281.

191 - T. MUIR - A theorem in continuants - Extension of a theorem in continuants with an important application - Phil. Mag., (5) 3 (1877).

192 - T. MUIR - On the phenomenon of greatest middle in the cycle of a class of a class of periodic continued fraction - Proc. Roy. Soc. Edinburgh, 12 (1884) 578-592.

193 - H. PADE - Mémoire sur les développements en fractions continues de la fonction exponentielle pouvant servir d'introduction à la théorie des fractions continues algébriques - Ann. Fac. Sci. de l'Ec. Norm. Sup., 16 (1899) 395-436.

194 - H. PADE - Sur la distribution des réduites anormales d'une fonction - C.R. Acad. Sc. Paris, 130 (1900).

195 - H. PADE - Sur le développement en fraction continue de la fonction F(h,ℓ,h',u) et la généralisation de la théorie des fonctions sphériques - C.R. Acad. Sc. Paris, 141 (1905) 819-821.

196 - O. PERRON - Über zwei Kettenbrüchen von H.S. Wall - Bayer, Akad. Wiss. Math. - Nat. Kl. S. - B., (1957) 1-13.

197 - O. PERRON - Die Lehre von dem Kettenbrüchen - Chelsea Pub. Co., New-York, 1950.

198 - O. PERRON - Über eine spezielle Klasse von Kettenbrüchen - Rend. Circ. Mat. Palermo, 29 (1910).

199 - S. PINCHERLE - Sur les fractions continues algébriques - Ann. Sci. Ec. Norm. Sup., (3) 6 (1889) 145-152.

200 - A. PRINGSHEIM - Über ein Konvergenzkriterium für Kettenbrüche mit positiven Gliedern - Sb. München, 29 (1899).

201 - E. ROUCHE - Mémoire sur le développement des fonctions en séries ordonnées suivant les dénominateurs des réduites d'une fraction continue - J. Ec. Polytechnique, 37 (1858).

202 - H. RUTISHAUSER - Über eine Verallgemeinerung der Kettenbrüche - Z. Angew. Math. Mech., 38 (1958) 278-279.

203 - H. RUTISHAUSER - Beschleunigung der Konvergenz einer gewisse Klasse von Kettenbrüchen - ZAMM, 38 (1958) 187.

204 - F.T. SCHUBERT - De transformatione series in fractionem continuam - Mem. Acad. Sc. Imp. St Pétersbourg, 7 (1815-1816).

205 - W.T. SCOTT - The corresponding continued fraction of a J-fraction - Ann. Math., (2) 51 (1950) 56-67.

206 - H. SIEBECK - Über periodischer Kettenbrüche - J. für die reine u. angew. math., 33 (1846) 71-77.

207 - V. SINGH, W.J.THRON - A family of best twin convergence regions for continued fractions - Proc. Amer. Math. Soc., 7 (1956) 277-282.

208 - J. SHERMAN - On the numerators of the convergents of the Stieltjes continued fraction - Tran. Amer. Math. Soc., 35 (1933) 64-87.

209 - T.J. STIELTJES - Sur la réduction en fraction continue d'une série précédant suivant les puissances descendantes de la variable - Ann. Fac. Sci. Toulouse, 3 (1889) 1-17.

210 - T.J. STIELTJES - Note sur quelques fractions continues - Quart. J. pure and appl. math., 25 (1891).

211 - T.J. STIELTJES - Recherches sur les fractions continues - Ann. Fac. Sci. Univ. Toulouse, 8 (1894) 1-122.

212 - T.J. STIELTJES - Sur un développement en fractions continues - C.R. Acad. Sc. Paris, 99 (1884) 508-509.

213 - T.J. STIELTJES - Recherches sur les fractions continues - Mémoires présen-

tés par divers savants à l'académie des Sciences de l'institut de France, Sciences et Mathématiques, (2) 32 (2) (1892).

214 - W.B. SWEEZY, W.J. THRON - Estimates of the speed of convergence of certain continued fractions - SIAM J. Numer. Anal., 4 (1967) 254-270.

215 - W.J. THRON - On parabolic convergence regions for continued fractions - Math. Zeit., 69 (1958) 172-182.

216 - W.J. THRON - Recent approaches to convergence theory of continued fractions - in "Padé approximants and their applications", P.R. Graves-Morris ed., Academic Press, New-York, 1973.

217 - W.J. THRON - A survey of recent convergence results for continued fractions - Rocky Mountains J. Math., 4 (1974) 273-282.

218 - W.J. THRON - Two families of twin convergence regions for continued fractions - Duke Math. J., 10 (1943) 677-685.

219 - W.J. THRON - Twin convergence regions for continued fractions - Amer. J. Math., 66 (1944) 428-439.

220 - W.J. THRON - A family of simple convergence regions for continued fractions - Duke Math. J., 11 (1944) 779-791.

221 - W.J. THRON - Convergence regions for the general continued fraction - Bull. Amer. Math. Soc., 49 (1943) 913-916.

222 - W.J. THRON - Some properties of the continued fraction $(1+d_o z)+K(z/(1+d_n z))$ - Bull. Amer. Math. Soc., 54 (1948) 206-218.

223 - W.J. THRON - Zwillingskonvergenzgebiete für Kettenbrüche $1+K(a_n/1)$, deren eines die Kreisscheibe $|a_{2n-1}| < \rho^2$ ist. - Math. Zeit., 70 (1959) 310-344.

224 - W.J. THRON - Convergence regions for continued fractions and other infinite processes - Amer. Math. Monthly, 68 (1961) 734-750.

225 - W.J. THRON - Convergence of sequences of linear fractional transformations and of continued fractions - J. Indian Math. Soc., 27 (1963) 103-127.

226 - W.J. THRON - Some results and problems in the analytic theory of continued fractions - Math. Student, 32 (1964) 61-73.

227 - W.J. THRON - On the convergence of the even part of certain continued fractions - Math. Zeit., 85 (1964) 268-273.

228 - J. TREMBLEY - Recherches sur les fractions continues - Mem. Acad. Roy. Sc. Belles-Let., Berlin (1794-1795)

229 - K.T. VAHLEN - Über Näherungswerte und Kettenbrüche - J. für die reine u. angew. math., 115 (1895) 221-233.

230 - E.B. VAN VLECK - On an extension of the 1894 memoir of Stieltjes - Trans. Amer. Math. Soc., 4 (1903) 297-332.

231 - E.B. VAN VLECK - On the convergence of continued fractions with complex elements - Trans. Amer. Math. Soc., 2 (1901) 215-233.

232 - E.B. VAN VLECK - On the convergence and character of the continued fraction

$a_1z/1+\ldots$ - Trans. Amer. Math. Soc., 2 (1901) 476-483.

233 - E.B. VAN VLECK - On the convergence of algebraic continued fractions whose coefficients have limiting values - Trans. Amer. Math. Soc., 5 (1904) 253-262.

234 - E.B. VAN VLECK - Selected topics in the theory of divergent series and continued fractions - Amer. Math. Soc. Colloquium Pub., 1, Boston Colloquium, 1903.

235 - E.B. VAN VLECK - On the convergence of the continued fraction of Gauss and other continued fractions - Ann. Math., 3 (1901) 1-18.

236 - H. VON KOCH - Quelques théorèmes concernant la théorie générale des fractions continues - Öfversigt af. Kongl. Vetenskaps - Akad. Förhandlingen, 52 (1895).

237 - B. VISCOVATOFF - De la méthode générale pour réduire toutes sortes de quantités en fractions continues - Mem. Acad. Sc. Imp. St Pétersbourg, 1 (1803-1804).

238 - H. WAADELAND - T-fractions from a different point of view - Rocky Mountains J. Math., 4 (1974) 391-394.

239 - H. WAADELAND - On T-fractions of functions holomorphic and bounded in a circular disc - Norske Vid. Selsk. Skr (Trondheim) 1964, n°8.

240 - H. WAADELAND - A convergence property of certain T-fraction expansions - Norske Vid. Selsk. Skr (Trondheim) 1966, n°9.

241 - H.S. WALL - On some criteria of Carleman for the complete convergence of a J-fraction - Bull. Amer. Math. Soc., 54 (1948) 528-532.

242 - H.S. WALL - Convergence of continued fractions in parabolic domains - Bull. Amer. Math. Soc., 55 (1949) 391-394.

243 - H.S. WALL - Note on a periodic continued fraction - Amer. Math. Monthly, 56 (1949) 96-97.

244 - H.S. WALL - Concerning continuous continued fractions and certain systems of Stieltjes integral equations - Rend. Circ. Mat. di Palermo, II,2 (1953) 73-84.

245 - H.S. WALL - Partially bounded continued fractions - Proc. Amer. Math. Soc., 7 (1956) 1090-1093.

246 - H.S. WALL - Some convergence problems for continued fractions - Amer. Math. Monthly, 54 (1957) 95-103.

247 - H.S. WALL - Note on a certain continued fraction - Bull. Amer. Math. Soc., 51 (1945) 930-934.

248 - H.S. WALL - Note on the expansion of a power series into a continued fraction - Bull. Amer. Math. Soc., 51 (1945) 97-105.

249 - H.S. WALL - Continued fraction expansions for functions with positive real parts - Bull. Amer. Math. Soc., 52 (1946) 138-143.

250 - H.S. WALL - Theorems on arbitrary J-fractions - Bull. Amer. Math. Soc., 52 (1946) 671-679.

251 - H.S. WALL - Bounded J-fractions - Bull. Amer. Math. Soc., 52 (1946) 686-693.

252 - H.S. WALL - Some recent developments in the theory of continued fractions - Bull. Amer. Math. Soc., 47 (1941) 405-423.

253 - H.S. WALL - A continued fraction related to some partition formulas of Euler - Amer. Math. Monthly, 48 (1941) 102-108.

254 - H.S. WALL - The behaviour of certain Stieltjes continued fractions near the singular line - Bull. Amer. Math. Soc., 48 (1942) 427-431.

255 - H.S. WALL - Continued fractions and bounded analytic functions - Bull. Amer. Math. Soc., 50 (1944) 110-119.

256 - H.S. WALL - Convergence criteria for continued fractions - Bull. Amer. Math. Soc., 17 (1931) 575-579.

257 - H.S. WALL - On the continued fractions which represent meromorphic functions - Bull. Amer. Math. Soc., 39 (1933) 942-952.

258 - H.S. WALL - Continued fractions and cross-ratios groups of Cremona transformations - Bull. Amer. Math. Soc., 40 (1934) 578-592.

259 - H.S. WALL - On continued fractions of the form $1+K_1^\infty(b_\nu z/1)$ - Bull. Amer. Math. Soc., 41 (1935) 727-736.

260 - H.S. WALL - On continued fractions representing functions - Bull. Amer. Math. Soc., 44 (1938) 94-99.

261 - H.S. WALL - Continued fractions and totally monotone sequences - Trans. Amer. Math. Soc., 48 (1940) 165-184.

262 - H.S. WALL - The analytic theory of continued fractions - Van Nostrand, New-York, 1948.

263 - H.S. WALL, J.J. DENNIS - The limit circle case for a positive definite J-fraction - Duke Math. J., 12 (1945) 255-273.

264 - H.S. WALL, H.L. GARABEDIAN - Hausdorff methods of summation and continued fractions - Trans. Amer. Math. Soc., 48 (1940) 185-207.

265 - H.S. WALL, E. HELLINGER - Contributions to the analytic theory of continued fractions and infinite matrices - Ann. Math., 44 (1943) 103-127.

266 - H.S. WALL, W. LEIGHTON - On the transformation and convergence of continued fractions - Amer. J. Math., 58 (1936) 267-281.

267 - H.S. WALL, J.F. PAYDON - The continued fraction as a sequence of linear transformations - Duke Math. J., 9 (1942) 360-372.

268 - H.S. WALL, R.E. LANE - Continued fractions with absolutely convergent even and odd parts - Trans. Amer. Math. Soc., 67 (1949) 368-380.

269 - H.S. WALL, W.T. SCOTT - On the convergence and divergence of continued fractions - Amer. J. Math., 69 (1947) 551-561.

270 - H.S. WALL, W.T. SCOTT - Continued fraction expansion for arbitrary power series - Ann. Math., 41 (1940) 325-349.

271 - H.S. WALL, W.T. SCOTT - Value regions for continued fractions - Bull. Amer. Math. Soc., 47 (1941) 580-585.

272 - H.S. WALL, W.T. SCOTT - Continued fractions - Natl. Math. Mag., 13 (1939) 1-18.

273 - H.S. WALL, W.T. SCOTT - A convergence theorem for continued fractions - Trans. Amer. Math. Soc., 47 (1940) 155-172.

274 - H.S. WALL, M. WETZEL - Quadratic forms and convergence regions for continued fractions - Duke Math. J., 11 (1944) 89-102.

275 - H.S. WALL, M. WETZEL - Contributions to the analytic theory of J-fractions - Trans. Amer. Math. Soc., 55 (1944) 373-397.

276 - R. WILSON - Divergent continued fractions and polar singularities - Proc. Lond. Math. Soc., 26 (1927) 159-168 / 27 (1928) 497-512 / 28 (1928) 128-144.

277 - R. WILSON - Divergent continued fractions and nonpolar singularities - Proc. Lond. Math. Soc., 30 (1928) 38-57.

278 - A. WINTNER - Ein qualitatives Kriterium für dis Konvergenz des assoziierten Kettenbrüche - Math. Zeit., 30 (1929) 285-289.

279 - L. WUYTACK - Extrapolation to the limit by using continued fraction interpolation - Rocky Mountains J. Math., 4 (1974) 395-397.

280 - P. WYNN - Continued fractions whose coefficients obey a noncommutative law of multiplication - Arch. Rat. Mech. Anal., 12 (1963) 273-312.

281 - P. WYNN - A note on the convergence of certain noncommutative continued fractions - MRC technical summary report 750, Madison, 1967.

282 - P. WYNN - Vector continued fractions - Linear Algebra, 1 (1968) 357-395.

283 - P. WYNN - Upon the definition of an integral as the limit of a continued fraction - Arch. Rat. Mech. Anal., 28 (1968) 83-148.

284 - P. WYNN - An arsenal of Algol procedures for the evaluation of continued fractions and for effecting the epsilon algorithm - Chiffres, 9 (1966) 327-362.

285 - P. WYNN - Four lectures on the numerical application of continued fractions - CIME summer school lectures, 1965.

286 - P. WYNN - Converging factors for continued fractions - Numer. Math., 1 (1959) 272-307 and 308-320.

287 - P. WYNN - A note on a method of Bradshaw for transforming slowly convergent series and continued fractions - Amer. Math. Monthly, 69 (1962) 883-889.

288 - P. WYNN - A numerical study of a result of Stieltjes - Chiffres, 6 (1963) 175-196.

289 - P. WYNN - On some recent developments in the theory and application of continued fractions - SIAM J. Numer. Anal., 1 (1964) 177-197.

290 - P. WYNN - Note on a converging factor for a certain continued fraction - Numer. Math., 5 (1963) 332-352.

291 - P. WYNN - The numerical efficiency of certain continued fraction expansions - Koninkl. Nederl. Akad. Wet., 65 A (1962) 127-148.

292 - P. WYNN - Complex numbers and other extensions to the Clifford algebra with on application to the theory of continued fractions, MRC technical summary report 646, Madison, 1966.

III - Applications

293 - A.C. AITKEN - On Bernoulli's numerical solution of algebraic equations - Proc. Roy. Soc. Edinburg, 46 (1926) 289-305.

294 - A.C. AITKEN - On interpolation by proportional parts, without use of differences - Proc. Edin. Math. Soc., 3 (1932) 56-76.

295 - R. ALT - Méthodes A-stables pour l'intégration des systèmes différentiels mal conditionnés - Thèse 3ème cycle, Paris, 1971.

296 - F.L. BAUER - Connections between the q-d algorithm of Rutishauser and the ε-algorithm of Wynn - Deutsche Forschungsgemeinschaft Tech. Rep. Ba/106, 1957.

297 - F.L. BAUER - Nonlinear sequence transformations - in "Approximation of functions", Garabedian ed., Elsevier, New-York, 1965.

298 - F.L. BAUER - The g-algorithm - SIAM J., 8 (1960) 1-17.

300 - M. BAUSSET - Une détermination des surfaces de choc par accélération de convergence - Colloque Euromech 58, Toulon, 12-14 mai 1975.

301 - L.C. BREAUX - A numerical study of the application of acceleration techniques and prediction algorithms to numerical integration - M. Sc. Thesis, Louisiana State Univ., New-Orleans, 1971.

302 - C. BREZINSKI - Convergence d'une forme confluente de l'ε-algorithme - C.R. Acad. Sc. Paris, 273 A (1971) 582-585.

303 - C. BREZINSKI - L'ε-algorithme et les suites totalement monotones et oscillantes - C.R. Acad. Sc. Paris, 276 A (1973) 305-308.

304 - C. BREZINSKI - Méthodes d'accélération de la convergence en analyse numérique - Thèse, Univ. de Grenoble, 1971.

305 - C. BREZINSKI - Application du ρ-algorithme à la quadrature numérique - C.R. Acad. Sc. Paris, 270 A (1970) 1252-1253.

306 - C. BREZINSKI - Etudes sur les ε et ρ-algorithmes - Numer. Math., 17 (1971) 153-162.

307 - C. BREZINSKI - Résultats sur les procédés de sommation et l'ε-algorithme - RIRO, R3 (1970) 147-153.

308 - C. BREZINSKI - Forme confluente de l'ε-algorithme topologique - Numer. Math., 23 (1975) 363-370.

309 - C. BREZINSKI - Review of methods to accelerate the convergence of sequences- Rend. Mat. Roma, 7 (1974) 303-316.

310 - C. BREZINSKI - Accélération de la convergence en analyse numérique - Publication 41, labo. de Calcul, Univ. de Lille, 1973.

311 - C. BREZINSKI - Conditions d'application et de convergence de procédés d'extrapolation - Numer. Math., 20 (1972) 64-79.

312 - C. BREZINSKI - Application de l'ε-algorithme à la résolution des systèmes non linéaires - C.R. Acad. Sc. Paris, 271 A (1970) 1174-1177.

313 - C. BREZINSKI - Intégration des systèmes différentiels à l'aide du ρ-algorithme - C.R. Acad. Sc. Paris, 278 A (1974) 875-878.

314 - C. BREZINSKI - Sur un algorithme de résolution des systèmes non linéaires - C.R. Acad. Sc. Paris, 272 A (1971) 145-148.

315 - C. BREZINSKI - Numerical stability of a quadratic method for solving systems of non linear equations - Computing, 14 (1975) 205-211.

316 - C. BREZINSKI - Computation of the eigenelements of a matrix by the ε-algorithm - Linear Algebra, 11 (1975) 7-20.

317 - C. BREZINSKI, M. CROUZEIX - Remarques sur le procédé Δ^2 d'Aitken - C.R. Acad. Sc. Paris, 270 A (1970) 896-898.

318 - C. BREZINSKI, A.C. RIEU - The solution of systems of equations using the ε-algorithm, and an application to boundary value problems - Math. Comp., 28 (1974) 731-741.

319 - H. CABANNES, M. BAUSSET - Application of the method of Padé to the determination of shock wavres - in "Problems of hydrodynamics and continuum mechanics ", in honor of L.I. Sedov, English ed. publ. by SIAM (1968) 95-114.

320 - H.K. CHENG, M. HAFEZ - Convergence acceleration of iterative solutions and transonic flow computations - Colloque Euromech 58, Toulon, 12-14 mai 1975.

321 - J.S.R. CHISHOLM - Padé approximants and linear integral equations - in "The Padé approximant in theoretical physics", G.A. Baker Jr. and J.L. Gammel eds., Academic Press, New-York, 1970.

322 - J.S.R. CHISHOLM - Application of Padé approximation to numerical integration - Rocky Mountains J. Math., 4 (1974) 159-168.

323 - J.S.R. CHISHOLM - Padé approximation of single variable integrals - Colloquium on computational methods in theoretical physics, Marseille, 1970.

324 - J.S.R. CHISHOLM - Accelerated convergence of sequences of quadrature approximants - second colloquium on computational methods in theoretical physics, Marseille, 1971.

325 - J.S.R. CHISHOLM, A.C. GENZ, G.E. ROWLANDS - Accelerated convergence of sequences of quadrature approximation - J. Comp. Phys., 10 (1972) 284-307.

326 - J. COUNTS, J.E. AKIN - The application of continued fractions to wave propagation problems in a viscoelastic rod - DEMVPI Research Rep. 1-1, Dept. of Eng. Mech., Virginia polytechnic institute, Blacksburg, 1968.

327 - J.D.P. DONNELLY - Applications of the q-d and ε-algorithms - in "Methods of numerical approximation", D.C. Handscomb ed., Pergamon Press, Oxford, 1965.

328 - B.L. EHLE - A-stable methods and Padé approximations to the exponential - SIAM J. Math. Anal., 4 (1973) 671-680.

329 - B.L. EHLE - On Padé approximations to the exponential function and A-stable methods for the numerical solution of initial value problems - Research rep. CSRR 2010, dept. of AACS, Univ. of Waterloo, Ontario, 1969.

330 - M. FROISSART - Applications of the Padé method to numerical analysis - Colloquium on computational methods in theoretical physics, Marseille, 1970.

331 - E. GEKELER - Über den ε-algorithmus von Wynn - Z. Angew. Math. Mech., 51 (1971) 53-54.

332 - E. GEKELER - On the solution of systems of equations by the epsilon algorithm of Wynn - Math. Comp., 26 (1972) 427-436.

333 - A. GENZ - The ε-algorithm and some other applications of Padé approximants in numerical analysis - in "Padé approximants", P.R. Graves- Morris ed., The institute of physics, London, 1973.

334 - A. GENZ - Applications of the ε-algorithm to quadrature problems - in "Padé approximants and their applications ", P.R. Graves-Morris ed., Academic Press, New-York, 1973.

335 - B. GERMAIN BONNE - Transformations de suites - RAIRO, R1 (1973) 84-90.

336 - M. HAFEZ, H.K. CHENG - On acceleration of convergence and shock-fitting in transonic flow computations - Univ. So. Calif. Memo., 1973.

337 - M. HAFEZ, H.K. CHENG - Convergence acceleration and shock fitting for transonic aerodynamics computations - AIAA paper, No. 75-51.

338 - P. HENRICI - Some applications of the quotient-difference algorithm - Proc. Symp. Appl. Math., 20 (1963) 159-183.

339 - P. HENRICI - The quotient-difference algorithm - NBS appl. Math. series, 49 (1958) 23-46.

340 - D.C. JOYCE - Survey of extrapolation processes in numerical analysis - SIAM Rev., 13 (1971) 435-490.

341 - D.K. KAHANER - Numerical quadrature by the ε-algorithm - Math. Comp., 26 (1972) 689-694.

342 - D. LEVIN - Development of non-linear transformations for improving convergence of sequences - Intern. J. Comp. Math., B3 (1973) 371-388.

343 - I.M. LONGMAN - Use of Padé table for approximate Laplace transform inversion - in "Padé approximants and their applications", P.R. Graves-Morris ed., Academic Press, New-York, 1973.

344 - E.D. MARTIN, H. LOMAX - Rapid finite difference computation of subsonic and transonic aerodynamic flows - AIAA paper, No. 74-11.

345 - I. MARX - Remark concerning a nonlinear sequence to sequence transformation - J. Math. Phys., 42 (1963) 334-335.

346 - K.J. OVERHOLT - Extended Aitken acceleration - BIT, 5 (1965) 122-132.

347 - R. PENNACCHI - Somma di serie numeriche mediante la trasformazione quadratica $T_{2,2}$ - Calcolo, 5 (1968) 51-61.

348 - R. PENNACCHI - La trasformazioni razionali di una successione - Calcolo, 5 (1968) 37-50.

349 - A. RONVEAUX - Padé approximant and homographic transformation of Riccati's phase equations - in "Padé approximants and thein applications", P.R. Graves-Morris ed., Academic Press, New-York, 1973.

350 - H. RUTISHAUSER - On a modification of the q-d algorithm with Graeffe type convergence - ZAMP, 13 (1962) 493-496.

351 - H. RUTISHAUSER - Bestimmung der Eigenwerte und Eigenvektoren einer Matrix mit Hilfe des Quotienten - Differenzen Algorithmus - ZAMP, 6 (1955) 387-401.

352 - H. RUTISHAUSER - Eine Formel von Wronski und ihre Bedentung für den Quotienten - Differenzen Algorithmus - ZAMP, 7 (1956) 164-169.

353 - H. RUTISHAUSER - Stabile Sonderfälle des Quotienten - Differenzen Algorithmus - Numer. Math., 5 (1963) 95-112.

354 - H. RUTISHAUSER - Der Quotienten - Differenzen Algorithmus - ZAMP, 5 (1954) 233-251.

355 - H. RUTISHAUSER - Anwendungen des Quotienten - Differenzen Algorithmus - ZAMP, 5 (1954) 496-508.

356 - J.R. SCHMIDT - On the numerical solution of linear simultaneous equations by an iterative method - Phil. Mag., 7 (1951) 369-383.

357 - L.W. SCHWARTZ - On the analytic structure of the Taylor-Maccall conical flow solution - ZAMP.

358 - L.W. SCHWARTZ - Series solution for the planar asymmetric blunt-body problem - Phys. Fluids.

359 - L.W. SCHWARTZ - Solutions to the asymmetric blunt-body problem using Padé approximants - Colloque Euromech 58, Toulon, 12-14 mai 1975.

360 - L.W. SCHWARTZ - Hypersonic flows generated by parabolic and paraboloidal shock waves - Phys. Fluids, 17 (1974) 1816-1821.

361 - R.E. SHAFER - On quadratic approximation - SIAM J. Numer. Anal., 11 (1974) 447-460.

362 - D. SHANKS - An analogy between transients and mathematical sequences and some nonlinear sequence to sequence transforms suggested by it - Naval Ordnance Lab. Mem. 9994, White Oak, Md., 1949.

363 - D. SHANKS - Non linear transformations of divergent and slowly convergent series - J. Math. Phys., 34 (1955) 1-42.

364 - M.D. VAN DYKE - Perturbation methods in fluid mechanics - Academic Press, New-York, 1964.

365 - M.D. VAN DYKE - Analysis and improvement of perburtation series - Quart. J. Mech. Appl. Math.

366 - A.H. VAN TUYL - Calculation of nozzle using Padé fractions - AIAA J., 11 (1973) 537-541.

367 - A.H. VAN TUYL - Application of methods for acceleration of convergence to the calculation of singularities of transonic flows - Colloque Euromech 58, Toulon, 12-14 mai 1975.

368 - A.H. VAN TUYL - The use of Padé fractions in the calculation of nozzle flows - Colloque Euromech 58, Toulon, 12-14 mai 1975.

369 - H.S. WALL, W.T. SCOTT - The transformation of series and sequences - Trans. Amer. Math. Soc., 51 (1942) 255-279.

370 - P.J.S. WATSON - Algorithms for differentiation and integration - in "Padé approximants and their applications", P.R. Graves-Morris ed., Academic Press, New-York, 1973.

371 - L. WUYTACK - The use of Padé approximation in numerical integration - Colloque Euromech 58, Toulon, 12-14 mai 1975.

372 - P. WYNN - On a connection between the first and the second confluent form of the ε-algorithm - Nieuw. Arch. Wisk., 11 (1963) 19-21.

373 - P. WYNN - Upon a second confluent form of the ε-algorithm - Proc. Glasgow Math. Soc., 5 (1962) 160-165.

374 - P. WYNN - A comparison between the numerical performances of the Euler transformation and the ε-algorithm - Chiffres, 1 (1961) 23-29.

375 - P. WYNN - On repeated application of the ε-algorithm - Chiffres, 4 (1961) 19-22.

376 - P. WYNN - The epsilon algorithm and operational formulas of numerical analysis - Math. Comp., 15 (1961) 151-158.

377 - P. WYNN - Acceleration techniques in numerical analysis with particular reference to problems in one independant variable - Proc. IFIP Congress, North Holland, (1962) 149-156.

378 - P. WYNN - A note on programming repeated application of the ε-algorithm - Chiffres, 8 (1965) 23-62.

379 - P. WYNN - The rational approximation of functions which are formally defined by a power series expansion - Math. Comp., 14 (1960) 147-186.

380 - P. WYNN - The abstract theory of the epsilon algorithm - Centre de recherches mathématiques n° 74, Univ. de Montréal, 1971.

381 - P. WYNN - Upon a hierarchy of epsilon arrays - Louisiana State Univ., New-Orleans, techn. rep. 46, 1970.

382 - P. WYNN - Invariants associated with the epsilon algorithm and its first confluent form - Rend. Circ. Mat. Palermo, (2) 21 (1972) 31-41.

383 - P. WYNN - Singular rules for certain nonlinear algorithms - BIT, 3 (1963) 175-195.

384 - P. WYNN - A sufficient condition for the instability of the ε-algorithm -

Nieuw. Arch. Wisk., 3 (1961) 117-119.

385 - P. WYNN - On the propagation of error in certain nonlinear algorithms - Numer. Math., 1 (1959) 142-149.

386 - P. WYNN - On the convergence and stability of the epsilon algorithm - SIAM J. Numer. Anal., 3 (1966) 91-122.

387 - P. WYNN - Hierarchies of arrays and function sequences associated with the epsilon algorithm and its first confluent form - Rend. Mat. Roma, 5 (1972) 819-852.

388 - P. WYNN - Accélération de la convergence de séries d'opérateurs en analyse numérique - C.R. Acad. Sc. Paris, 276 A (1973) 803-806.

389 - P. WYNN - Transformations de séries à l'aide de l'ε-algorithme - C.R. Acad. Sc. Paris, 275 A (1972) 1351-1353.

390 - P. WYNN - Upon some continuous prediction algorithms - Calcolo, 9 (1972) 197-234 and 235-278.

391 - P. WYNN - Confluent forms of certain nonlinear algorithms - Arch. Math., 11 (1960) 223-234.

392 - P. WYNN - A note on a confluent form of the ε-algorithm - Arch. Math., 11 (1960) 237-240.

393 - P. WYNN - On a procrustean technique for the numerical transformation of slowly convergent sequences and series - Proc. Camb. Phil. Soc., 52 (1956) 663-671.

394 - P. WYNN - Transformations to accelerate the convergence of Fourier series - Gertrude Blanch anniversary vol., Wright Patterson Air Force Base, (1967) 339-379.

395 - P. WYNN - Partial differential equations associated with certain nonlinear algorithms - ZAMP, 15 (1964) 273-289.

396 - P. WYNN - A numerical method for estimating parameters in mathematical models - Centre de recherches mathématiques, Univ. de Montréal, rep. CRM - 443, 1974.

397 - P. WYNN - On a device for computing the $e_m(S_n)$ transformation - MTAC, 10 (1956) 91-96.

398 - P. WYNN - A convergence theory of some methods of integration - Centre de recherches mathématiques, Univ. de Montréal, rep. CRM-193, 1972.

399 - P. WYNN - A sufficient condition for the instability of the q-d algorithm - Numer. Math., 1 (1959) 203-207.

400 - P. WYNN - A comparison technique for the numerical transformation of slowly convergent series based on the use of rational functions - Numer. Math., 4 (1962) 8-14.

IV - MISCELLANEOUS

401 - G.D. ALLEN, C.K. CHUI, W.R. MADYCH, F.J. NARCOWICH, P.W. SMITH - Padé approximation of Stieltjes series - J. Appr. Theory, 14 (1975) 302-316.

402 - G.D. ALLEN, C.K. CHUI, W.R. MADYCH, F.J. NARCOWICH, P.W. SMITH - Padé approximants, Nuttall's formula and a maximum principle - to appear.

403 - M.F. BARNSLEY - Correction terms for Padé approximants - J. Math. Phys., 16 (1975) 918-928

404 - C. BREZINSKI - Matrices semi-définies positives et suites de moments - to appear.

405 - F. CORDELLIER - Une nouvelle forme de l'ε-algorithme et ses variantes - Colloque Euromech 58, Toulon, 12-14 mai 1975.

406 - F. CORDELLIER - Règles particulières pour l'ε-algorithme vectoriel - Colloque d'Analyse Numérique, La Grande Motte, mai 1975.

407 - F. CORDELLIER - Interprétation géométrique d'une étape de l'ε-algorithme - Publ. 40, Labo. de Calcul, Univ. de Lille, 1973.

408 - F. CORDELLIER - L'ε-algorithme vectoriel : interprétation géométrique et règles singulières - Colloque d'analyse numérique, Gourette, mai 1974.

409 - A.C. GENZ - The approximate calculation of multidimensional integrals using extrapolation methods - Ph. D. Thesis, Univ. of Kent at Canterbury, 1975.

410 - P. HILLION - Méthode d'Aitken itérée pour les suites oscillantes d'approximations - C.R. Acad. Sc. Paris, 280 A (1975) 1701-1704.

411 - S.T. PENG, A. HESSEL - Convergence of noncommutative continued fractions - SIAM J. Math. Anal., 6 (1975) 724-727.

412 - J. TODD - The lemniscate constants - Comm. ACM, 18 (1975) 14-19.

UER d'IEEA - Informatique
BP 36
59650 - Villeneuve d'Ascq
FRANCE

Lecture Notes in Physics

Bisher erschienen/Already published

Vol. 1: J. C. Erdmann, Wärmeleitung in Kristallen, theoretische Grundlagen und fortgeschrittene experimentelle Methoden. II, 283 Seiten. 1969.

Vol. 2: K. Hepp, Théorie de la renormalisation. III, 215 pages. 1969.

Vol. 3: A. Martin, Scattering Theory: Unitarity, Analyticity and Crossing. IV, 125 pages. 1969.

Vol. 4: G. Ludwig, Deutung des Begriffs „physikalische Theorie" und axiomatische Grundlegung der Hilbertraumstruktur der Quantenmechanik durch Hauptsätze des Messens. 1970. Vergriffen.

Vol. 5: M. Schaaf, The Reduction of the Product of Two Irreducible Unitary Representations of the Proper Orthochronous Quantummechanical Poincare Group. IV, 120 pages. 1970.

Vol. 6: Group Representations in Mathematics and Physics. Edited by V. Bargmann. V, 340 pages. 1970.

Vol. 7: R. Balescu, J. L. Lebowitz, I. Prigogine, P. Résibois, Z. W. Salsburg, Lectures in Statistical Physics. V, 181 pages. 1971.

Vol. 8· Proceedings of the Second International Conference on Numerical Methods in Fluid Dynamics. Edited by M. Holt. 1971. Out of print.

Vol. 9: D. W. Robinson, The Thermodynamic Pressure in Quantum Statistical Mechanics. V, 115 pages. 1971.

Vol. 10: J. M. Stewart, Non-Equilibrium-Relativistic Kinetic Theory. III, 113 pages. 1971.

Vol. 11: O. Steinmann, Perturbation Expansions in Axiomatic Field Theory. III, 126 pages. 1971.

Vol. 12: Statistical Models and Turbulence. Edited by C. Van Atta and M. Rosenblatt. Reprint of the First Edition. VIII, 492 pages. 1975.

Vol. 13: M. Ryan, Hamiltonian Cosmology. VII, 169 pages. 1972.

Vol. 14: Methods of Local and Global Differential Geometry in General Relativity. Edited by D. Farnsworth, J. Fink, J. Porter and A. Thompson. V, 188 pages.

Vol. 15: M. Fierz, Vorlesungen zur Entwicklungsgeschichte der Mechanik. V, 97 Seiten. 1972.

Vol. 16: H.-O. Georgii, Phasenübergang 1. Art bei Gittergasmodellen. IX, 167 Seiten. 1972.

Vol. 17: Strong Interaction Physics. Edited by W. Rühl and A. Vancura. V, 405 pages. 1973.

Vol. 18: Proceedings of the Third International Conference on Numerical Methods in Fluid Mechanics, Vol. I. Edited by H. Cabannes and R. Temam. VII, 186 pages. 1973.

Vol. 19: Proceedings of the Third International Conference on Nemerical Methods in Fluid Mechanics, Vol. II. Edited by H. Cabannes and R. Temam. VII, 275 pages. 1973.

Vol. 20: Statistical Mechanics and Mathematical Problems. Edited by A. Lenard. VIII, 247 pages. 1973.

Vol. 21: Optimization and Stability Problems in Continuum Mechanics. Edited by P. K. C. Wang. V, 94 pages. 1973.

Vol. 22: Proceedings of the Europhysics Study Conference on Intermediate Processes in Nuclear Reactions. Edited by N. Cindro, P. Kulišić and Th. Mayer-Kuckuk. XIV, 329 pages. 1973.

Vol. 23: Nuclear Structure Physics. Proceedings 1973. Edited by U. Smilansky, I. Talmi, and H. A. Weidenmüller. XII, 296 pages. 1973.

Vol. 24: R. F. Snipes, Statistical Mechanical Theory of the Electrolytic Transport of Non-electrolytes. V, 210 pages. 1973.

Vol. 25: Constructive Quantum Field Theory. The 1973 "Ettore Majorana" International School of Mathematical Physics. Edited by G. Velo and A. Wightman. III, 331 pages. 1973.

Vol. 26: A. Hubert, Theorie der Domänenwände in geordneten Medien. XII, 377 Seiten. 1974.

Vol. 27: R. K. Zeytounian, Notes sur les Ecoulements Rotationnels de Fluides Parfaits. XIII, 407 pages. 1974.

Vol. 28: Lectures in Statistical Physics. Edited by W. C. Schieve and J. S. Turner. V. 342 pages. 1974.

Vol. 29: Foundations of Quantum Mechanics and Ordered Linear Spaces. Advanced Study Institute, Marburg 1973. Edited by A. Hartkämper and H. Neumann. VI, 355 pages. 1974.

Vol. 30: Polarization Nuclear Physics. Proceedings 1973. Edited by D. Fick. IX, 292 pages. 1974.

Vol. 31: Transport Phenomena. Sitges International Schools of Statistical Mechanics, June 1974. Edited by G. Kirczenow and J. Marro. XIV, 517 pages. 1974.

Vol. 32: Particles, Quantum Fields and Statistical Mechanics. Proceedings 1973. Edited by M. Alexanian and A. Zepeda. V, 132 pages. 1975.

Vol. 33: Classical and Quantum Mechanical Aspects of Heavy Ion Collisions. Proceedings 1974. Edited by H. L. Harney, P. Braun-Munzinger, and C. K. Gelbke. VII, 311 pages. 1975.

Vol. 34: One-Dimensional Conductors. GPS Summer School Proceedings, 1974. Edited by H. G. Schuster. VII, 371 pages. 1975.

Vol. 35: Proceedings of the Fourth International Conference on Numerical Methods in Fluid Dynamics. June 24–28, 1974, University of Colorado. Edited by R. D. Richtmyer. V, 457 pages. 1975.

Vol. 36: R. Gatignol, Théorie Cinétique des Gaz à Répartition Discrète de Vitesses. II, 219 pages. 1975.

Vol. 37: Trends in Elementary Particle Theory. Proceedings 1974. Edited by H. Rollnik and K. Dietz. V, 472 pages. 1975.

Vol. 38: Dynamical Systems, Theory and Applications. Proceedings 1974. Edited by J. Moser. VI, 624 pages. 1975.

Vol. 39: International Symposium on Mathematical Problems in Theoretical Physics. Proceedings 1975. Edited by H. Araki. XII, 562 pages. 1975.

Vol. 40: Effective Interactions and Operators in Nuclei. Proceedings 1975. Edited by B. R. Barrett. XII, 339 pages. 1975.

Vol. 41: Progress in Numerical Fluid Dynamics. Proceedings 1974. Edited by H. J. Wirz. V, 471 pages. 1975.

Vol. 42: H II Regions and Related Topics. Proceedings 1975. Edited by D. Downes and T. L. Wilson. XII, 488 pages. 1975.

Vol. 43: Laser Spectroscopy. Proceedings 1975. Edited by S. Haroche, J. C. Pebay-Peyroula, T. W. Hänsch, and S. E. Harris. X, 466 pages. 1975.

Vol. 44: R. A. Breuer, Gravitational Perturbation Theory and Synchrotron Radiation. VI, 196 pages. 1975.

Vol. 45: Dynamical Concepts on Scaling Violation and the New Resonances in e^+e^- Annihilation. Edited by B. Humpert. VII, 248 pages. 1976.

Vol. 46: E. J. Flaherty, Hermitian and Kählerian Geometry in Relativity. VIII. 365 pages. 1976.

Vol. 47: Padé Approximants Method and Its Applications to Mechanics. Edited by H. Cabannes. XV, 267 pages. 1976

SPRINGER TRACTS IN MODERN PHYSICS

Ergebnisse der exakten Naturwissenschaften

Springer-Verlag
Berlin
Heidelberg
New York

Volume 66

30 figures. III, 173 pages. 1973
ISBN 3-540-06189-4

Quantum Statistics

in Optics and Solid-State Physics

R. Graham: Statistical Theory of Instabilities in Stationary Nonequilibrium Systems with Applications to Lasers and Nonlinear Optics.
F. Haake: Statistical Treatment of Open Systems by Generalized Master Equations.

Volume 67

III, 69 pages. 1973
ISBN 3-540-06216-5

S. Ferrara, R. Gatto, A. F. Grillo:

Conformal Algebra in Space-Time

and Operator Product Expansion

Introduction to the Conformal Group in Space-Time. Broken Conformal Symmetry. Restrictions from Conformal Covariance on Equal-Time Commutators. Manifestly Conformal Covariant Structure of Space-Time. Conformal Invariant Vacuum Expectation Values. Operator Products and Conformal Invariance on the Light-Cone. Consequences of Exact Conformal Symmetry on Operator Product Expansions. Conclusions and Outlook.

Volume 68

77 figures. 48 tables. III, 205 pages. 1973
ISBN 3-540-06341-2

Solid-State Physics

D. Schmid: Nuclear Magnetic Double Resonance — Principles and Applications in Solid-State Physics.
D. Bäuerle: Vibrational Spectra of Electron and Hydrogen Centers in Ionic Crystals.
J. Behringer: Factor Group Analysis Revisited and Unified.

Volume 69

13 figures. III, 121 pages. 1973
ISBN 3-540-06376-5

Astrophysics

G. Börner: On the Properties of Matter in Neutron Stars.
J. Stewart, M. Walker: Black Holes: the Outside Story.

Volume 70

II, 135 pages. 1974
ISBN 3-540-06630-6

Quantum Optics

G. S. Agarwal: Quantum Statistical Theories of Spontaneous Emission and their Relation to Other Approaches.

Volume 71

116 figures. III, 245 pages. 1974
ISBN 3-540-06641-1

Nuclear Physics

H. Überall: Study of Nuclear Structure by Muon Capture.
P. Singer: Emission of Particles Following Muon Capture in Intermediate and Heavy Nuclei.
J. S. Levinger: The Two and Three Body Problem.

Volume 72

32 figures. II, 145 pages. 1974
ISBN 3-540-06742-6

D. Langbein:

Theory of Van der Waals Attraction

Introduction. Pair Interactions. Multiplet Interactions. Macroscopic Particles. Retardation. Retarded Dispersion Energy. Schrödinger Formalism. Electrons and Photons.

Volume 73

110 figures. VI, 303 pages. 1975
ISBN 3-540-06943-7

Excitons at High Density

Editors: H. Haken, S. Nikitine
Biexcitons. Electron-Hole Droplets. Biexcitons and Droplets. Special Optical Properties of Excitons at High Density. Laser Action of Excitons. Excitonic Polaritons at Higher Densities.

Volume 74

75 figures. III, 153 pages. 1974
ISBN 3-540-06946-1

Solid-State Physics

G. Bauer: Determination of Electron Temperatures and of Hot Electron Distribution Functions in Semiconductors.
G. Borstel, H. J. Falge, A. Otto: Surface and Bulk Phonon-Polaritons Observed by Attenuated Total Reflection.

Selected Issues from

Lecture Notes in Mathematics